PIG MANAGEMENT AND PRODUCTION
A Practical Guide for Farmers and Students

Southam King David 24th. Mr. Lionel Organ's senior stock boar—Sire of Supreme Champion Female and First Prize and Reserve Male Champion at N.P.B.A. Sale, Edinburgh 1971—and to date has sired 24 M.L.C. Performance tested boars with ratings of over 120 points.

PIG MANAGEMENT AND PRODUCTION

A practical guide for farmers and students

Derek H. Goodwin
N.D.A., S.C.D.A., F.T.C. (Agric.), Cert. Ed. (Birmingham)

Lecturer in Animal Husbandry
Gloucestershire College of Agriculture

HUTCHINSON EDUCATIONAL

HUTCHINSON EDUCATIONAL LTD
3 Fitzroy Square, London W1

London Melbourne Sydney Auckland
Wellington Johannesburg Cape Town
and agencies throughout the world

First published 1973

*This book has been set in Imprint type, printed in Great Britain
on smooth wove paper by Anchor Press, and
bound by Wm. Brendon, both of Tiptree, Essex*

ISBN 0 09 110890 x (cased)
     0 09 110891 8 (paper)

Acknowledgements

In writing this book I have drawn freely on the writings of research scientists. Any book on livestock production must be based on the research work of the late John Hammond and his colleagues at Cambridge. The reader should certainly refer to Hammond's *Physiology of Farm Animals*, and C. P. McMeekan's *Principles of Livestock Production*.

I am extremely grateful to Dr. A. Eden, A.D.A.S., for advice on nutrition and permission to use Tables 6 and 7. Dr. D. R. Melrose, M.L.C., for papers on artificial insemination in pigs. Dr. R. Braude, N.I.R.D., for papers on cage rearing pigs. Bill Marshall, B.O.C.M./Silcock, for help with pig housing. The Meat and Livestock Commission for permission to use Tables 1, 12, and 13 and for providing plates by Ian Clook and Jack Fisher. The Agricultural Research Council for permission to use Table 8 from *The Nutrient Requirements of Farm Livestock, No. 3 Pigs*. The National Pig Breeders' Association for Table 4, and the British Landrace Pig Society for Table 5. John Shepherd and Company for plate 'Cage rearing' and *The Insulation Hand Book* for special permission to use Tables 14 and 15. H.M.S.O. for Tables 3 and 9.

My special thanks are due to my friends and colleagues in Gloucestershire for providing material and help with the manuscript: Christopher Rogers, A.D.A.S., Colin Hawkes-Wood, M.L.C., Norman Kelbrick, technical librarian. Lionel Organ for permission to use Table 22. Stuart Freeman, who took the castration photographs and others from which many drawings were taken. Gordon Cinderby, who patiently read the entire manuscript and proofs, and lastly my wife for drawing the 'roughs' and Mrs. Janice Wilson for typing the manuscript.

5

The author would like to express his gratitude to the following for their help in granting permission to reproduce material during the collection of illustrations and data for this book:

Mr. Lionel Organ of Manor Farm, Southam, Cheltenham, Gloucestershire; John Shepherd and Sons; the editorial director of *Insulation Handbook*; and the staffs of the Agricultural Development and Advisory Service; the Ministry of Agriculture Cambridge; the National Institute for Research in Dairying, Reading; the British Landrace Pig Society; and the National Pig Breeders' Association.

Contents

7

11

13

14

Preface

The aim of this book is to explain simply and attractively the basic principles of pig production, and to provide a comprehensive guide to modern practice. The book is written primarily for students, past and present, of the county agricultural colleges, but should also be of interest to farmers, pigmen and apprentices.

During the past decade tremendous changes have taken place in the pig industry, due, partly, to economic pressures, but mainly through the enormous increase in scientific information. Pig breeding, for example, has improved to such a degree that one can hardly recognise pictures of pigs taken twenty years ago and present-day pigs as being of the same breed! Nutrition is no longer dealt with by hit or miss on the farm rations, but rather designed, costed and calculated by computers. Housing also has seen many changes from the cottager's sty to the sow stall.

It is under these conditions that the young apprentice, or indeed the practising pig farmer, may find himself rather confused and bewildered with the various systems of production and their permutations that go together to produce such a standard commodity as the present-day bacon pig!

It is hoped, therefore, that this book will assist the beginner to analyse much of the current information avialable to him, and from this discover the fundamental principles on which he may build his knowledge into sound farming practice.

Author's Note

In view of the Council of Technical Examining Bodies decision to recommend that all agricultural examinations should use S.I. units as from 1973, I have used metric units throughout the text. But, as it is likely that imperial units will still be used by the general public for some time yet, I have included conversion tables in the form of a quick Ready Reckoner in the Appendix.

One Pig Products and Market Requirements

Pigs are kept solely in this country for the production of meat, and it is claimed that pigs produce more edible dishes than any other type of livestock. Pig meat in its fresh state is called pork; when it is processed or 'cured' it is called bacon.

When a pig is slaughtered it produces a carcass, which consists of bone, muscle (red meat) and fat. It is usual to include the head, tail, and feet as part of the carcass. The internal organs are removed to provide edible offals and the waste products, such as the skin, hair, gut contents, blood, and hooves, are called inedible offals.

Pork is produced from young quickly grown pigs when they are about $3\frac{1}{2}$–4 months old. The pork carcass is divided into two sides, each of which in turn is cut into six main joints, namely the leg of pork, hind loin, fore loin, spare-rib, hand and belly. The leg is usually sold as a roasting joint, whilst the loins may either be roasted or cut into pork chops for frying. The spare-rib and belly pork are cheaper joints, which may be roasted, fried, or used for making pork pies or sausage.

Pork is light in colour when compared with other red meats, and has a fresh moist appearance. The housewife looks for a 'pinky' colour, and a large proportion of lean meat to bone and fat. The fat should have a smooth appearance, and be firm and white in colour.

Pork consumption has more than doubled since 1939, and now stands at around 12 kg per annum per head of the population. This has been brought about largely by overcoming the old prejudice of only eating pork when there was an *R* in the month, and with the help of the household refrigerator. Fresh pork and pork products are often termed *convenience foods*, meaning that they can be quickly prepared by busy housewives who are in full or part-time employment.

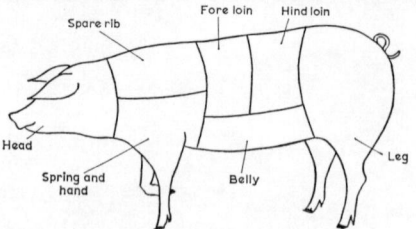

Fig. 1.   Pork pig

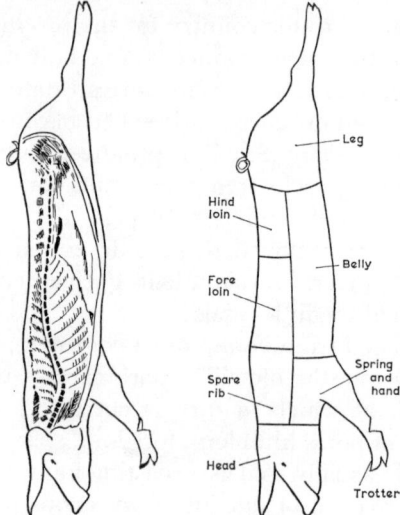

Fig. 2.   Pork carcass

On the farm, pork production has many advantages, in that a quick turnover of capital is made, porkers are more efficient converters of food than baconers or heavy hogs, and that there is no grading scheme for the end products, although butchers, of course, pay extra for good-quality pigs. The most sought-after porkers are those which are of the white-bacon breeds, which are lean but well fleshed and weigh around 45–55 kg live-weight. Some butchers, however, prefer a heavier pig of 60–70 kg liveweight. These are called cutters.

Cutters are used for both the fresh meat and manufacturing trade. The ribs and loins yield excellent chops when the surplus

20

fat is trimmed. The legs may be sold as pork and the fore-end manufactured into pies and sausage.

Bacon is produced by curing fresh pigmeat with sodium chloride (common salt) and sodium nitrate (saltpetre).

It is interesting to note that bacon has been produced since about 200 B.C., when the Romans salted down pigmeat according to a recipe of Cato, a Roman butcher. Cato's cure is still used today by farmers who wish to 'home cure' pigs on the farm.

The majority of bacon is produced by the *Wiltshire cure*, which was introduced in 1877 by *Mr. Harris* of Calne, Wiltshire. The *Wiltshire cure* requires pigs of about 90–95 kg liveweight.

Bacon carcass

The main aim in quality bacon production is to breed and fatten pigs that will yield a carcass with a high proportion of lean meat to fat and bone; and additionally, to produce a maximum of the cuts that command the highest retail price, such as the loin and ham.

We can consider the bacon pig as three main cuts, namely the *fore-end* or shoulder, the *middle* which includes the ribs and loin region, and the *ham* or *gammon* which is the hind leg.

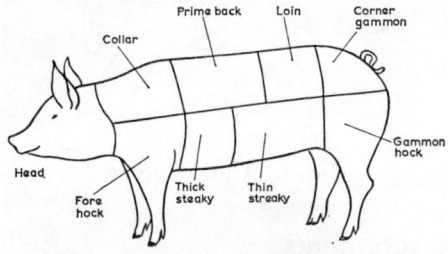

Fig. 3.   Bacon pig

The fore-end produces the cheaper cuts because the meat is of lower quality and is usually used as boiling joints. The middle produces the rashers of bacon, and it is desirable that this should be as long as possible in order to produce the maximum number of rashers. The ham or gammon is the most valuable cut, producing high-quality meat suitable for grilling as 'gammon steak' or, after boiling, for slicing as cold ham.

21

The quality of the carcass determines the price that the producer receives and all carcasses are graded according to their length of side and the thickness of backfat. An optical probe is used to determine the backfat measurement over the shoulder, mid-back and loin.

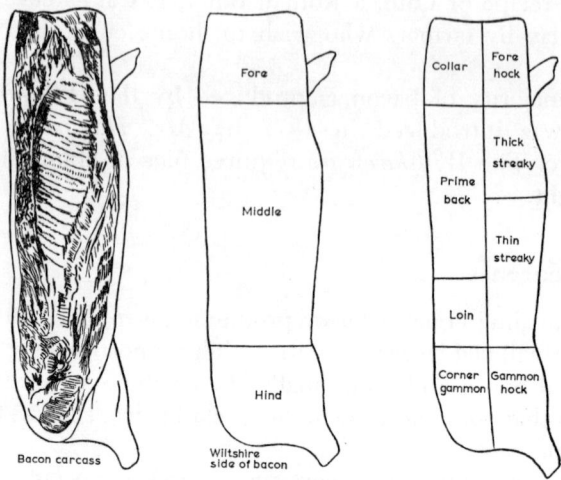

Fig. 4.   Bacon carcass

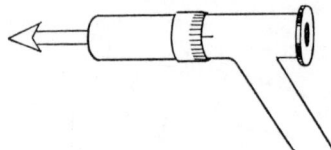

Fig. 5.   Optical probe

C and K measurements

In order to assess the lean meat content of the carcass, the optical probe measurements are taken in line with the head of the last rib. The probe is inserted at points 4·5 cm (P1); 6·5 cm (P2) and 8 cm (P3) from the centre of the back. (See Fig. 5).

The operator is then able to read off the measurements in millimetres recorded on the dial of the probe. By using a special formula it is possible to calculate the amount of 'eye muscle' in the rasher.

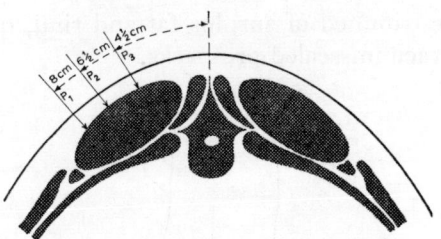

Fig. 6. C and K measurements

The length of carcass is measured from the first rib to the pelvic bone. Good carcass will measure around 775–800 mm between these points. The carcass is then graded into the categories 1, 2, or C, depending upon the length, C and K measurements at shoulder, loin and mid-back, and weight of carcass.

Table 1. Bacon carcass grading standards

Grade	Carcass weight	Length (minimum)	Backfat shoulder/loin	'P2' Probe measurement (maximum)
1	61 kg to 70 kg	775 mm	45/25 mm	22 mm
2	61 kg to 70 kg	775 mm	48/28 mm	22 mm
	or	775 mm	45/25 mm	24 mm
	or	—	45/25 mm	22 mm
C	59 kg to 77 kg	—	49/29 mm or over	—

Heavy hogs

There is an increasing demand for heavy pigs of around 120 kg for manufacturing purposes. In recent years the processing industry has developed new techniques that enable the heavy carcass to be used for pork, bacon, and manufacturing purposes. The hind legs may be used as fresh joints or cured for ham; the middles are used for bacon, and the shoulders and belly manufactured into tinned shoulder, pork pies or sausage. The heavy hogs offer greater flexibility to the industry in that they make it possible to meet the fluctuating demands of the general public. Furthermore, the bacon produced from these

23

carcasses are trimmed of surplus fat and rind, quickly cured, and sold in vacuum-sealed pre-packs.

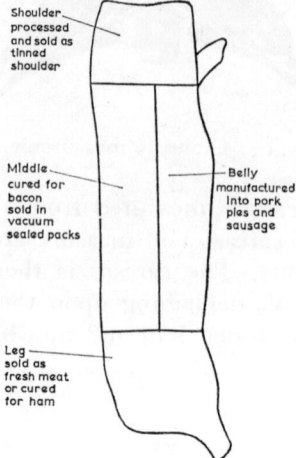

Fig. 7. Heavy-hog carcass

Young boar meat

There is abundant evidence to prove that young boar pigs can produce up to 30% more lean meat than the castrate when slaughtered at similar weights, and that the entire is more efficient in utilising food, and grows more quickly than the castrate.

Unfortunately, boar meat contains a substance which produces an objectionable odour when the meat is cooked. This is less apparent with very young boars, but definitely increases with age, and in mature boars is quite pronounced.

If in the future scientists are able to devise methods of eliminating boar taint, then undoubtedly we shall keep entire pigs for bacon production.

At the present time there is considerable interest in boar meat for pork, where the taint is hardly noticeable.

Fat sows

Fat sows are manufactured into pork pies, lard and sausage, the meat being unsuitable for bacon. Good-quality fat sows command a relatively high price compared with heavy hogs. This is due to the excellent dressing percentage, and because the meat is claimed to give the sausages more flavour owing to its age.

Edible offals consist of the brains, blood, trachea (windpipe), lungs, heart, liver, spleen, stomach, intestines and internal fat. Secondary offals are the head and trotters. The lungs are used in the preparation of faggots. The liver, heart and spleen are sold as 'pig's fry'. Trotters are sold for boiling, and the waste fat is rendered down as lard. Blood is used in making 'black pudding'.

Inedible offals are the skin, hair, bones and waste products. Pigskin is extremely valuable, being a hard-wearing leather which is capable of taking a very good polish. It is used for making shoes, riding saddles, ladies' handbags, gloves, briefcases and wallets. Pig hair is used for upholstery and making brushes. The waste products, blood, faeces and urine, are used as manure to restore and maintain soil fertility.

Supply and demand

Table 2. Meat consumption in the United Kingdom

	kg per person per annum
Beef and veal	21
Mutton and lamb	11·5
Fresh pigmeat	8·5
Bacon	11·5
Offal	4
Poultry	9·5

The national pig herd has increased in size more than four-fold since the Second World War and today stands at around eight million pigs. The greatest increase in consumption has been in the fresh-meat and tinned-meat industries, bacon consumption remaining relatively stable. Home production accounts for approximately 50% of the total for pig products, but only one-third of the bacon market, Denmark being the largest supplier of imported bacon.

Changes in the industry

The general structure of pig production has changed markedly in the past decade and considerable changes are likely to occur in future years. The number of pig farmers is falling, but the size of herds is enlarging, particularly with breeding stock. Today there are numerous herds of over 100 sows and many herds of 500, whilst occasionally one hears of the 2000-sow herd.

It is likely that this trend will continue, which should lead to greater stability in the industry, with less farmers controlling much larger units.

Table 3.   Growth of herd size (England and Wales)

	1961	1962	1963	1964	1965	1966
Average number of sows on holding	7·4	8·2	8·8	9·5	10·1	10·7
Average number of pigs on holding	41·6	46·9	51·0	57·7	65·7	70·7

Source: Ministry of Agriculture, Fisheries and Food

Two Reproduction in Pigs

Female reproductive organs

The general construction of the sow's reproductive organs is shown in Fig. 8. These consist of the two *ovaries* which produce eggs or ova; the fallopian tubes, or oviducts which convey the eggs after being released; the *uterus* or womb in which the eggs

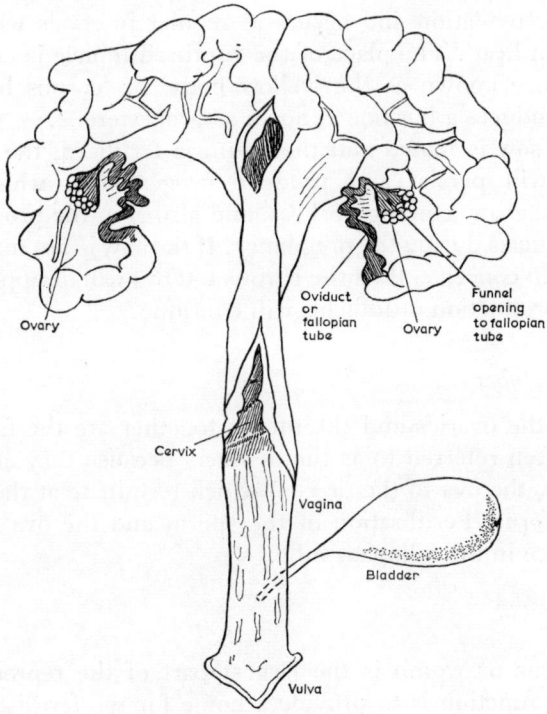

Fig. 8.   Female reproductive organs

are implanted, and where the embryo will grow and develop; the *cervix* which forms a strong muscular collar that opens and closes the mouth of the uterus; the *vagina* which is a fairly large tube which connects the uterus to the *vulva* which is the external opening to the reproductive organs.

Ovaries

The sow has two ovaries which are suspended by ligaments in the loin region of the body and are small rounded bodies, measuring about 10–25 mm across. The ovaries produce eggs or ova and various hormones which assist in reproduction and also affect changes in body growth.

The ova (eggs) are produced in sac-like structures called the Graffian follicles. The follicles also produce a fluid, so when the ovum is fully developed and the follicle ruptures, the fluid carries the ovum from the ovary to the fallopian tube in readiness for fertilisation by the male sperm. This process is known as ovulation and occurs at regular intervals when the sow is 'on heat'. The place of the ruptured follicle is taken by a substance known as the 'yellow body' or 'corpus luteum', which produces a hormone known as progesterone.

If the sow is mated and the ovum is fertilised, the corpus luteum will persist and produce progesterone which will prevent the sow coming on heat, and also stop the production of the follicles during the pregnancy. If the sow is not mated, or she fails to conceive, then the corpus luteum will disappear and normal production of follicles will continue.

Fallopian tubes

Linking the ovaries and the uterus together are the fallopian tubes (often referred to as the oviducts because they are used to convey the ova to the uterus) which terminate at the horns of the uterus. Fertilisation of the sperm and the ova usually takes place in the fallopian tube.

Uterus

The uterus or womb is the largest part of the reproductive tract. Its function is to provide a home for the fertilised eggs until they develop into a fully grown litter of piglets. In pigs this 'growing period' is approximately 114–116 days and known as the 'gestation period'.

Immediately after the ova is fertilised by the male sperm the egg begins to grow and at the same time passes down the fallopian tube into the uterus. The egg attaches itself to the wall of the uterus; a thin 'skin' or 'placenta' is then formed around the growing egg which is now called the embryo. The placenta then produces 'buttons' or cotyledons, which attach themselves to the wall of the uterus, and through which food is passed to the growing pig via the navel cord or umbilicus.

It can be seen, therefore, that the foetus is not connected directly to the sow, but is carried and fed inside the placenta.

Most sows shed at each ovulation period (heat) about twenty eggs which are fertilised, but some of them start to atrophy or degenerate during the early stages of pregnancy (see page 97). Occasionally, a mummified foetus will be farrowed with the rest of the pigs.

At birth or parturition the placenta is broken and the piglet delivered without this covering. Later the placenta, now called the 'afterbirth' or cleansing, is discharged from the uterus and the sow is said to have cleansed.

Cervix

This is a strong muscular collar which joins the vagina to the uterus. It is sometimes referred to as the neck of the uterus and takes the form of a constricted passage, the walls of which are thrown into folds or ridges. In the act of mating the corkscrew-like tip of the boar's penis is 'locked' in the muscular folds of the cervix and this stimulates the boar to ejaculate.

Vagina

The vagina is a fairly large tube which connects the uterus to the external opening called the vulva. There is a small opening in it for the urinary tract from the bladder. In the act of mating the boar's penis enters the vagina in order to deposit semen in the uterus.

Vulva

The vulva is the external opening of the tract. It consists of two lateral lips or labia, and at the lower end is found the *clitoris*, a small rod-like structure. When the sow is 'on heat' the vulva becomes swollen and in white breeds turns a 'pinky' colour.

Male reproductive organs

The essential parts of the male reproductive organs are the two testicles or *testes* which produce the sperms; the two *vas deferens* or *spermatic cords* down which the sperms travel; three sets of glands which produce fluid substances to form the semen in which the sperms swim; the urethra, which is a passage through which both urine and sperms pass; and the penis, which when protruded is used for implanting the sperms in the female during mating or *service*.

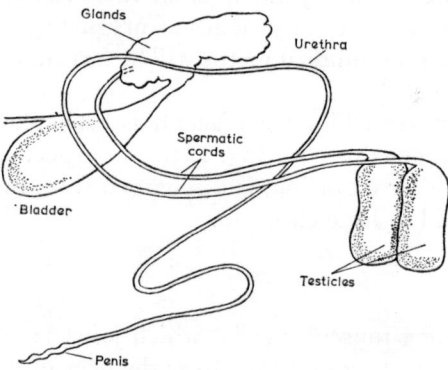

Fig. 9. Male reproductive organs

Testicles

The boar, like other male farm animals, has two testicles which are suspended outside the body, but enclosed in a sac or scrotum. The normal temperature of the testicles is about 0·1°C less than the body; the scrotum is outside the body in order to regulate the temperature. In cold weather the scrotum contracts, thus bringing the testicles nearer the body and in hot conditions the scrotum expands. It is possible in some males that the testicles are found in the abdominal cavity and do not descend into the scrotum, when this happens, or if only one testicle descends, the pig is known as a rig. Occasionally a pig is found with part of the gut in the scrotum, this is referred to as a rupture, or *scrotal hernia*.

30

The function of the testicles is to produce live sperms, which are single free-moving cells that are capable of uniting with the female ovum to produce a fertilised egg. Also the testicles produce the male sex hormone *testosterone*, which is responsible for inducing the desire to mate and the secondary sex characteristics. Thus the entire male will be masculine in appearance, whereas if the pig is castrated (surgical removal of the testicles) the castrate will have a feminine appearance and show no interest in mating.

Sperms

Large numbers of sperms are produced in the testicles in the *semin ferous tubules*. The sperms pass from the tubules into a coiled tube attached to the rear of the testicle and called the *epididymis*. The sperms are stored for a short period in the epididymis, where they mature and then they pass into the vas deferens, which is a duct leading back into the body cavity and connects to the urethra.

As the sperms move along the vas deferens or spermatic cords they mix with fluids produced in the 'accessory' glands, namely the prostrate gland, the seminal vesicle, and Cowper's gland. These fluids mix with the sperms and act as a carrier. The fluid is now called semen. The boar produces approximately 200–250 cm³ of semen containing about 50–60 000 million sperm at each ejaculation.

Urethra and penis

The two vas deferens or spermatic cords join together to form the urethra, which is a long tube that connects the bladder to the penis and therefore acts as a common vessel for both urine and semen. The penis is the final part of the reproductive tract and is normally retained in a loose fold of skin—the prepuce or sheath. It is formed of *erectile tissue* which contains spaces which can be filled with blood. In the act of mating the penis becomes filled with blood so that it is rigid and can penetrate the female vagina.

Artificial insemination—A.I.

Artificial insemination, or A.I., is a means of collecting natural semen from a boar and implanting it in the vagina or uterus of the sow. There are many advantages to be gained from A.I., for example it can allow the use of top-quality boars in small herds which could not justify keeping an individual boar. It is invaluable as a means of controlling health by reducing the risk of transmitting certain diseases as in natural service.

Pig insemination

In some parts of the country A.I. operators visit the farm and inseminate the sows, similar to the cattle A.I. service. However, at present the majority of inseminations are carried out by the pigman or his local veterinary surgeon. The semen is obtained from the A.I. centre by means of postal or railway service. To inseminate a sow the following equipment is required:

1 A catheter, which is a long rubber tube, similar to the boar's penis (Fig. 10).
2 A thin walled plastic container which holds the semen and is attached to the catheter.
3 A small quantity of liquid paraffin.

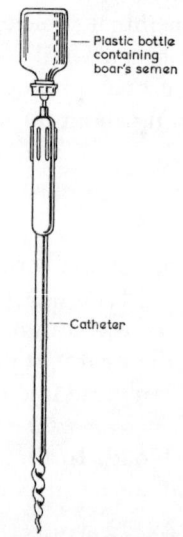

Fig. 10.   Catheter

The sow is normally 'on heat' for two and a half days, During this period and in the absence of a boar she will stand

32

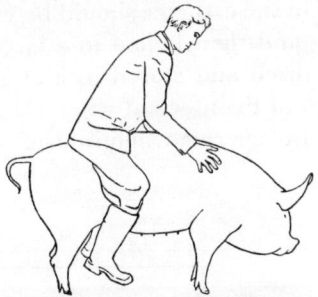

Fig. 11.   Testing gilt for *AI*

firmly when pressure is applied to her loin region. (See Fig. 11) This 'standing period' lasts for up to 29 hours and the effective period for insemination is 12–30 hours after the onset of oestrous, that is when the sow will first stand to the boar.

The catheter should first be lubricated with a few drops of liquid paraffin and then inserted carefully and firmly upwards into the vagina. Once inserted, the catheter should be twisted gently, anti-clockwise, until it 'locks' in the cervix. Once in the correct position, the plastic semen bottle should be attached and squeezed very gently to allow the semen to flow. The sow will take about 10–15 minutes to contract all the semen. On no account should the operator try to hurry the procedure by squeezing the bottle too hard, or the semen will be forced back out of the vulva.

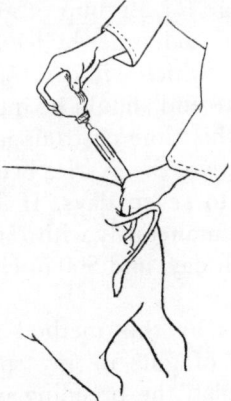

Fig. 12.   Artificial insemination

33

After insemination the catheter should be washed thoroughly in clean, hot water, and then boiled in a large saucepan for at least ten minutes, dried and stored in a clean polythene bag. Never use detergent or disinfectant when cleaning the catheter, as this will destroy the sperms when in use.

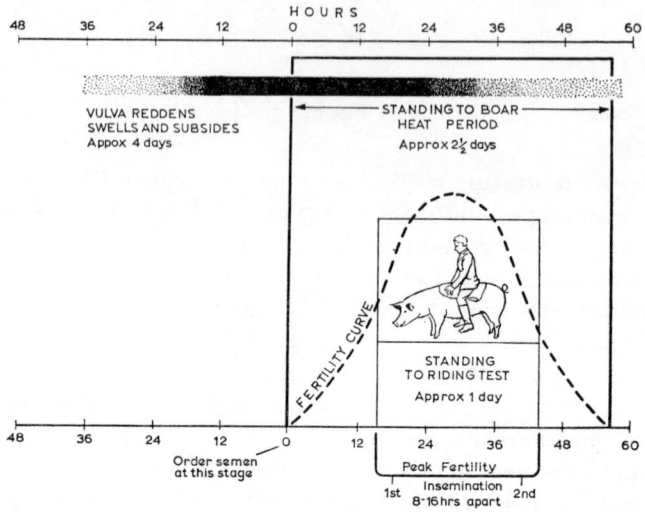

Fig. 13. The vital time

Controlled oestrous

It is possible to arrange for a group of gilts to come 'on heat' all within a day or so of each other by feeding a chemical added to their normal rations, which will control the heat period. The chemical is methalibure and should be included in the food for twenty days. During this time oestrous is inhibited, but, once the chemical is withdrawn, the ovaries become active and 'heat' occurs in about four to seven days. If A.I. is used the gilts should be injected subcutaneously with 750 iu of Pregnant mare serum on the twentieth day, and 500 iu H.C.G. on the twenty-fifth day.

Controlling oestrous by this method makes it possible to 'batch farrow' groups of gilts in any specified number, and allows the farmer to plan the breeding programme in such a way as to make the optimum use of existing buildings. In herds

34

where artificial insemination is used there is the added advantage that the A.I. operator can inseminate several gilts during one visit to the farm.

References
GROVES, T. W., *Veterinary Record* (1966).
SELLERS, K. C., *British Farmer* (1968).
PALGE, C., DAY, B. N., and GROVES, T. W., *Veterinary Record* (1968).

Three Growth and Development

Hammond defined growth as 'the increase in liveweight of an animal until a mature size is reached'. Development is defined as 'change in body shape or conformation'. Development also includes changes in body structure—for example, development of mammary tissue in the pregnant female.

Growth before birth

The pig's life begins at conception, with the union of the boar's sperm and the sow's egg. Each parent will contribute substantially one-half to the inheritance of the resulting offspring, although, of course, there may be some variation in the influence of certain sows and boars. In the early stages of pregnancy the developing pigs (foetuses) will grow quite slowly, and are surrounded in the uterus by large amounts of fluids and tissue. These fluids act as a 'cushion' to the foetuses and thereby protect them from injury should the sow be bullied by others or injured in some way.

As the pregnancy advances, the foetuses make rapid growth in size and weight, their bulk replacing much of the fluids. In fact we can say that two-thirds of the growth will take place during the last one-third of the pregnancy. It is therefore important that the sow should receive adequate nutrition at this stage, although care should be taken not to overfeed a sow just prior to farrowing, as this may cause 'overstocking' of the udder.

Birth weight

The size of the offspring at birth is controlled by factors other than nutrition. The more important of these are hereditary, and the number of piglets in the litter. A small litter of, say, six pigs will produce heavier piglets at birth than a larger litter of twelve or fourteen, but the total weight of the large litter will probably exceed that of the smaller number.

Birth weight is also controlled by the sow, by means of a special substance in the bloodstream which prevents the foetus from growing to such a size as to prevent parturition. Were it not for this control, the mating of large breeds like the Large White Boar to a small breed like the Berkshire would give rise to difficulties at birth. These occur only in exceptional circumstances. There is at present considerable interest amongst farmers and research workers into the effect of heavy birth weight on subsequent performance. Some farmers will only select future breeding gilts from those which weighed over 1·5 kg at birth.

Growth after birth

Provided the sow produces ample supplies of milk and the management is sound, baby pigs are capable of rapid growth in their early life; for example, a pig weighing 1 kg at birth will double that weight in its first week and quadruple it in the first three weeks!

If, however, we plot the weekly live-weight gain against the age of the pig, we shall see a moderate live-weight gain (about 1·5 kg per week) in the early stages, followed by rapid growth (4–4·5 kg w.l.w.g.) until puberty is reached, and then a downward trend (2–3 kg w.l.w.g.) as the pig approaches maturity.

McMeeken has shown the growth rate in farm animals as an S or sigmoid curve when plotting the live-weight gain against the age of the animals (see Fig. 14).

It should be noted that when the pig is growing fastest—the steepest part of the S curve—it is also the most efficient in converting food into live-weight gain. This is referred to as good *food conversion ratio* (f.c.r.)—which is the ratio between food eaten and the increase in body weight.

As the pig approaches maturity, the food conversion ratio will widen, that is, it will take more food to increase the body weight. We must also remember that growth is influenced by heredity (i.e. the breed and strain of pig), the environment in which the pig is kept and freedom from disease. The inheritance

37

of an animal sets the upper limit to its possible growth and its final size, whilst the nutrition, environment, management, and disease factors will determine its actual rate of growth and mature size.

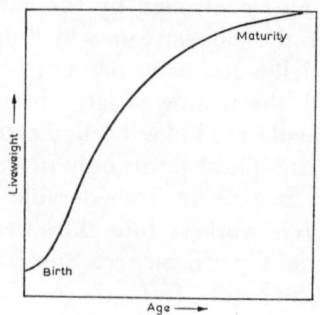

Fig. 14.   McMeekan's growth curve

Development—changes in body proportions

At birth the baby pig has a relatively large head, the legs are long and the body is small. The fully grown pig, on the other hand, has a comparatively small head, the legs are short in proportion to the body.

Changes in body shape are due to different parts growing at different rates. The head and legs (skeleton) are early to develop, while the body, particularly the loin and ham, are the last parts to reach a mature size.

The body, which later becomes the carcass, consists of three main tissues: *bone*, which forms the skeleton, *muscle*, which is the red meat, and *fat*. These three tissues grow in a very definite order. Bone is the first to develop, muscle is intermediate but tends to follow bone fairly closely, while fat is the last to develop and grows fastest as the pig approaches maturity.

The young animal, therefore, contains a higher proportion of bone and muscle, and a lower percentage of fat than the fully mature animal.

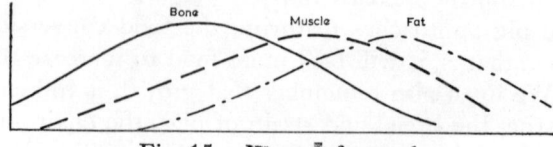

Fig. 15.   Waves of growth

38

Early maturity

Early maturing breeds such as the Berkshire and Middle White are those in which the 'waves of growth' are steep and follow each other closely. Such breeds fatten at light weights and early ages. Thus the early maturing breed will have a similar carcass (proportion of bone–muscle and fat) at around 45 kg as a late maturing breed at 90–100 kg.

Whilst in theory the early-maturing breed is ideal for pork production, in practice we find a marked preference, by the consumer, for well-fleshed, late-maturing breeds, which when slaughtered before maturity possess very little fat but rather a high porportion of bone; these are the most popular type of pigs for the pork market.

Late maturity

Late-maturing breeds, which we call the bacon breeds, are those in which the 'waves of growth' are less steep and the pigs mature at a heavier weight than the early-maturing breeds.

The current practice in bacon production is to feed bacon breeds on a high plane of nutrition until they reach 45–50 kg l.w., to promote rapid and efficient growth, and then to restrict food until the pig reaches 90–95 kg at slaughter weight. Thus by killing the pigs before they reach maturity and carefully managing the pigs' diet, bacon pigs with a high lean meat content are produced. This is known as the high–low plane of nutrition.

The influence of nutrition

The ratio of lean meat to bone and fat in the carcass can be greatly influenced by the quantity and quality of the food given to the pig during its lifetime. This was demonstrated by *Dr. McMeekan* in 1940, when he published the results of his experiments on feeding four groups of similarly bred pigs on four different planes of nutrition.

The groups were fed as follows:
High–High plane—meaning that the pigs received adequate amounts of food.
High–Low plane—the pigs received adequate feed for the first sixteen weeks and were then restricted or rationed.

Low–High plane—in this case the food was restricted until the pigs reached sixteen weeks and then fed adequate amounts.

Low–Low plane—the pigs were restricted to receiving only sufficient food to allow a slow growth rate.

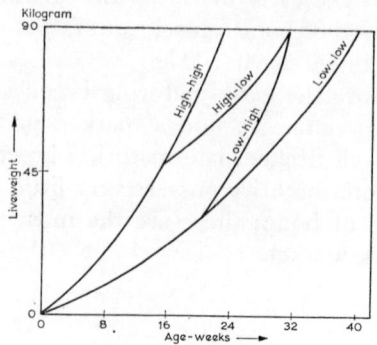

Fig. 16.   McMeekan's feeding experiment

When the pigs were slaughtered it was found that the *High–High* group, which had grown quickly, possessed a high lean meat content, but also had excessive fat.

The *High–Low* group produced excellent carcasses with a high proportion of lean meat and small amount of fat. The high plane of nutrition in early life had produced maximum growth of bone and muscle, whilst the restricted feeding from sixteen weeks had limited the deposit of fat.

The *Low–High* group resulted in poor growth rate in the early stages and then in the final stages large amounts of fat were deposited over a comparatively small amount of muscle.

The *Low–Low* group produced inferior carcasses, with a large amount of bone, and a relatively poor development of the later maturing parts such as the loin and hams, or in butchers' terms they were 'underfinished'.

Although it is over thirty years since McMeekan published this work, we still find that the majority of our pigs respond to this type of management. However, in recent years many breeders have developed strains of pure-bred pigs, and hybrid pigs, that will produce good carcasses even when fed on an unrestricted diet. Scientists also have now produced much better rations than those used by McMeekan, and this too can greatly affect the final carcass.

40

Changes in internal organs

In addition to the changes that take place in the body, similar 'waves of growth' occur in the internal organs. The main organs like the brain, heart, lungs and digestive system are early developing, whilst the reproductive organs and udder tissue in the female do not develop until much later.

Offals and dressing percentage

When the pig is slaughtered most of the internal organs are utilised as offal. Young animals have a higher proportion of offals in relation to the carcass weight than will the older animal. This explains why pork pigs have a lower dressing percentage or carcass yield than older stock.

	Dressing percentage
Pork	70–72
Cutters	72–73
Bacon	73–75
Heavy hogs	75–80
Fat sows	Over 80

References

HAMMOND, John, *Farm Animals,* Edward Arnold (1940).

HAMMOND, John, *Progress in the Physiology of Farm Animals,* Butterworth (1954–5).

McMEEKAN, C. P., *Principles of Animal Production,* Whitcombe and Tombs, New Zealand (1943).

YEATES, N. T. M., *Modern Aspects of Animal Production,* Butterworth, (1965).

Four Pig Breeds

There are nine main pig breeds recognised in the United Kingdom and recently several 'new' breeds have been imported from Canada, America and Belgium. The breeds may be conveniently classified as bacon, dual-purpose and pork.

Classification of pig breeds		
Bacon	*Dual-purpose*	*Pork*
Large White	British Saddleback	Middle White
Landrace	Large Black	Berkshire
Welsh	Gloucestershire Old Spot	
Tamworth		

Recently imported breeds

Hampshire, Lacombe, Pietrain, Duroc Jersey, Poland China.

The bacon breeds are by far the most important, and in fact produce the majority of pork today, as well as bacon. Bacon breeds are late maturing, quick growing, and produce carcasses full of lean meat when slaughtered before they mature. Bacon breeds are also more prolific and better mothers than pork breeds, which is one of the chief reasons for their popularity.

The dual-purpose breeds, particularly the Saddlebacks, are popular where an outdoor pig enterprise is required, the coloured breeds being more hardy and thrifty than the white breeds.

The true pork breeds which were popular a few years ago have lost favour recently, this being due mainly to their low breeding performance and their carcasses tending to be a little too fat for present-day demand.

42

Large White

This breed is numerically the most important pig breed in this country. The present-day Large White breed can produce ideal bacon pigs. In fact it would be difficult to fault some of the best strains. It originated in Yorkshire, and is often referred to as the Yorkshire.

The Large White is easily recognised by its slightly dished face and erect ears. The breed standard requires a pure white skin and coat, the body is long and deep, and yields carcasses with a high proportion of lean meat to bone and fat. The pigs grow quickly, reaching bacon weight in 180 days or less, and convert food efficiently.

The sows are prolific and good mothers, although in the past some strains were found to be rather clumsy. Large White boars are used extensively for top crossing with dual-purpose breeds to provide the well-known 'Blue' pigs which are capable of very rapid growth.

Today many farmers cross the Large White with the Landrace or Welsh to gain the advantages of hybrid vigour. The breed has been exported to many countries (Herd Book—N.P.B.A.).

British Landrace

The Landrace breed was imported from Sweden in 1949 and in 1953. It is descended from the Danish Landrace—a breed renowned for excellent carcass quality.

The British Landrace is now well established in the U.K. and is numerically the second most important breed. In appearance it is similar to the Welsh, being entirely white skinned and with lop ears. The sows are good mothers and prolific, their progeny making well-fleshed, long, lean carcasses. The boars are frequently used for 'top crossing' with other breeds where a white sire is required. The breed has been exported to countries all over the world (Herd Book—British Landrace Pig Society).

Welsh

Originating in Wales, the Welsh breed has always been recognised as a hardy breed and able to thrive either when kept indoors or outdoors. In recent years the breed has been greatly improved in its ability to produce high-quality carcasses with economical food conversion.

The breed is very similar in appearance to the Landrace, and indeed some strains carry a 'dash' of Landrace blood. The sows are excellent mothers, extremely prolific and deep milkers. Welsh pigs are suitable for all trades and can be called a multi-purpose breed (Herd Book—N.P.B.A.).

Tamworth

The Tamworth is one of the older British breeds. The Tam-worth is recognised by its golden-brown coat and is often referred to as a Sandy pig. The breed produces top-quality meat, but unfortunately the pigs tend to grow rather slowly. They are a 'hardy' breed and do well out of doors.

The sows are not very prolific, and tend to be poor mothers. Consequently the breed has lost favour in the U.K. in recent years, although it is held in considerable esteem in the United States of America and Canada. In New Zealand the Berkshire × Tamworth pig is very popular (Herd Book—N.P.B.A.).

Dual-purpose breeds

British Saddleback

The British Saddleback breed was formed in 1967 by the fusion of the Essex and Wessex Saddleback. Thus to date the 'new breed' has the advantages of hybrid vigour and the performance figures are most encouraging—N.P.B.A. figures show 9·27 pigs reared per recorded litter.

British Saddlebacks are easily recognised by their lop ears, black hair, and white band running over the shoulders and down the front legs.

The breed is hardy and the sows are often kept out of doors.

Many breeders of 'hybrid' pigs use the Saddleback dam in their breeding programme (Herd Book—N.P.B.A.).

Pietrain

In 1964 eighty-four Pietrain pigs were imported into the U.K. from Belgium by the Pig Industry Development Authority. (Now integrated Meat and Livestock Commission.) The purpose of the importation was to test the breed under controlled conditions, and compare the results with British breeds. The trial period ended in April 1971 and now the Pietrain is available.

In appearance the breed has a white skin with black patches and is lop eared; compared with native breeds the Pietrain is less prolific, grows more slowly and has a smaller appetite but similar feed-conversion efficiency. The carcass, however, contains more lean and a larger eye muscle. When slaughtered it also has a higher killing-out percentage than native breeds.

Although it would appear unlikely that as a pure breed the Pietrain has little to offer over our native breeds, it may well become a constituent in three-or-four-way crossing or in the formation of new synthetic lines.

The National Pig Breeders' Association has assumed responsibility for the breed and has formed a Pietrain Committee within the Association. In 1971 there were over 400 Pietrain breeding stock in the United Kingdom.

Hampshire

The Hampshire breed, which is similar in appearance to British Saddleback except that it has erect ears, was imported from America in 1962 and 1967. After extensive testing to compare its performance with existing British breeds, the Hampshire was released from control by the Ministry of Agriculture in March 1970. As a pure breed the Hampshire is not as prolific as our native breeds, but when crossed with Large Whites the carcasses consistently produced more lean meat and a better eye muscle than pure-bred Large Whites. The carcasses were, however, somewhat shorter and the backfat measurements were sometimes greater.

It would appear that like the Pietrain, the Hampshire has a contribution to make in cross-breeding and the formation of synthetic lines.

The breed is one of the constituent N.P.B.A. breeds and has its own Breed Council.

Large Black

This breed originated in East Anglia about 100 years ago. Pigs

45

of the earliest type were very heavy, fat and rather coarse. Present-day strains are much improved, and produce very good pigs when crossed with a white-bacon breed. The sows are excellent mothers, hardy, prolific and possess good milking qualities. They do very well when kept outdoors.

The breed has been widely exported. It is particularly popular in West Germany, where it is known as the Cornwall. In New Zealand it is called the Devon (Herd Book—N.P.B.A.).

Gloucestershire Old Spot

This is a white breed with black patches and spots in the skin. It is very hardy, and does well out of doors; the sows are good mothers, milk well and are noted for their long breeding lives. Like the other dual-purpose breeds, they are usually crossed with a white-bacon boar; the progeny is well suited for pork and heavy hog production (Herd Book—N.P.B.A.).

Lacombe

Introduced from Canada in the 1950's the Lacombe breed is also being used as a constituent of synthetic lines. It is a fairly large pig with white skin and somewhat resembles a Large White × Landrace in appearance. It is included in the hybrid for its rapid liveweight gain.

Other imported breeds

At the time of writing there are several other imported breeds undergoing tests in this country, some may well be released in the near future.

These include the Poland China, Duroc Jersey and American Yorkshire.

Pork breeds

Berkshire

The Berkshire is a black breed with a white flash over a dished face, erect ears and white markings on the feet. The pigs are very compact, mature early and produce carcasses with a high

46

ratio of valuable cuts. The sows, unfortunately, are not very prolific, which is the main reason for their decline in popularity. However, the breed is very popular in New Zealand, where it is crossed with the Tamworth and Large White. The Berkshire × Large White is known as the New Zealand National Hybrid (Herd Book—N.P.B.A.).

Middle White

The Middle White has a similar origin to the Large White. It is an early-maturing breed kept primarily for pork production. It is much shorter and thicker than the Large White, and is recognised by its white coat and dished face. Like the Berkshire, the breed has fallen out of fashion in recent years, due mainly to it being less prolific than other breeds (Herd Book—N.P.B.A.).

Table 4

Litter records

Details of litters notified to the N.P.B.A. in 1970 with corresponding figures for the three preceding years are given in this following table.

Breed	1970	1969	1968	1967
Berkshire	30	36	43	62
British Saddleback	1599	1727	2173	1251
G.O.S.	141	125	102	118
Hampshire	346	—	—	—
Landrace	187	121	143	200
Large Black	240	238	317	307
Large White	22 428	25 103	27 298	28 400
Middle White	49	51	51	44
Tamworth	26	36	37	52
Welsh	6282	6186	6246	6185
Wessex Saddleback	—	—	—	745
Cross-Bred	1285	1727	1612	1144
Special Registers	951	300	—	—
	33 564	35 650	38 022	38 508

Registrations

Details of registrations during 1970 and 1969 will be found in the following table:

Breed	1970 Boars	1970 Sows	1970 Total	1969 Boars	1969 Sows	1969 Total
Berkshire	14	34	48	11	27	38
British Saddleback	187	887	1074	194	856	1050
G.O.S.	13	81	94	14	79	93
Hampshire	126	356	482	—	—	—
Landrace	20	59	79	13	64	77
Large Black	57	217	274	66	207	273
Large White	3553	11 066	14 619	3409	10 279	13 688
Middle White	11	19	30	11	26	37
Tamworth	10	33	43	11	26	37
Welsh	817	2 980	3797	780	2872	3652
Special Registers	262	704	966	188	394	582
	5070	16 436	21 506	4697	14 830	19 527

Table 5. *Details of pigs notified to the British Landrace Pig Society: 1967–70*

Birth Notification Cards	
1967	19 500
1968	18 326
1969	16 575
1970	16 018

Herd Book registrations			
	Boars	*Gilts*	*Total*
1967	2368	8369	10 737
1968	2150	7103	9253
1969	2119	7355	9474
1970	2152	7834	9986

Five The Hybrid Pig-breeding Industry

Probably the most misused word in animal breeding is the term *hybrid*. The dictionary definition is 'offspring of two animals or plants of different species'. This, of course, implies cross-breeding or indeed mongrel or mixed breeding. Hybrid can also mean mating a donkey with a horse to produce the hybrid mule, or stallion mated to an ass to produce a hinny.

It might, therefore, be worth defining what we mean by a hybrid pig and hybrid vigour before we discuss the merits of hybrid breeding.

The Hybrid pig

Can be defined as a cross between two or more selected strains or breeds of pigs of known ancestry and performance. This process of hybridisation or cross-breeding generally results in improved litter performance characteristics in the hybrid progeny when compared with either of the pure-bred parents. In the production of modern hybrid strains of pig, not only are the litter traits important, but also growth rate, food conversion, and carcass quality. Improvement of these characteristics are achieved by selection of the pure-bred lines from which the parents of the hybrids are drawn. Different pure-bred lines will be stronger in certain traits and, therefore, the success of the hybrid pig will depend not only on the hybridisation but also on the careful selection of the desired characteristics in the pure-bred stock, and the combination of these lines, in the hybrid breeding programme. Thus if we have a strain of pigs that is known to produce good carcasses, and a further strain that grow quickly and convert their food efficiently, by mating the two together we can produce a hybrid with the trait for carcass

49

quality and economy of production. In practice we usually find three or more strains of pigs are used to produce the final hybrid, for example:

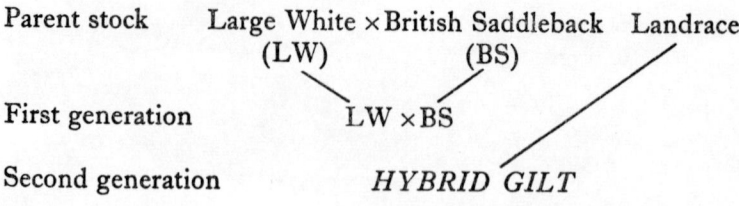

Fig. 17.   Breeding hybrids

If we cross a Large White boar from a strain known to have *rapid growth rate* with a strain of British Saddleback sows known to be *prolific* and *deep milkers* we shall produce first-generation gilts of known performance for these traits. The LW ×BS gilts can now be mated to a Landrace boar—known to possess the trait for *carcass quality*—and the offspring or second generation will be the hybrid gilts specially suited for bacon production when further crossed with a white boar.

Hybrid vigour

Hybrid vigour can be defined as the superior performance of the offspring over the average performance of the parents that occurs when two different breeds are mated together: for example if a Large White boar, from a strain known to produce heavy weaners, is mated to a similar type of Landrace sow, the progeny will not, as one might expect, be of similar weight to the parent stock. They will in fact be heavier. This increase in performance is called hybrid vigour.

Considerable research into pig breeding has shown that certain traits are more affected by hybrid vigour than others. Generally we find that cross-bred sows are more prolific and milk better than either of their parent pure-bred families. Similar research has also shown that the litters from cross-bred sows tend to be heavier at three weeks of age.

These, then, are some of the reasons for producing hybrid pigs. In general we can expect larger litters, heavier weaning

50

weights, lower mortality rates, better food conversion, faster growth rate, and more uniform quality in the carcass, with the specialist hybrid pig than with ordinary commercial pure-breds.

Breeding hybrids

There are many large commercial firms and pig farmers supplying first-quality hybrid gilts for sale. It can be argued that the large organisations, with their expert knowledge, computers, statisticians and geneticists, can supply better breeding stock more economically than the average farmer and certainly today hybrid pig breeding is a thriving industry.

The breeding programme of such enterprises can be illustrated by the following diagram

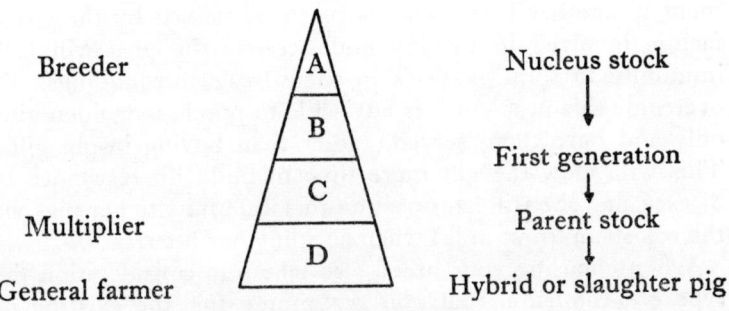

Fig. 18. Production of hybrids

The breeder or commercial firm maintains the pure strains (A) and may use, for example, the Hampshire for lean-meat content; Saddlebacks for fecundity; and Large White or Landrace for food conversion—quick growth rate etc. These strains are selected by the geneticist and the breeding programme is carefully formulated, to produce the first generation (B). Selected gilts are then sent to other farms (C) to multiply the numbers, and the resulting progeny are sold as the *such and such name hybrid*. The commercial farmer then buys in maiden hybrid gilts, mates them to a pure-bred boar, and the resulting progeny (D) will be sold for slaughter.

51

Choosing your breed

Pedigree, crossbred or hybrid?

The newcomer to pig keeping may well be bewildered by the number of pure breeds, crossbreds and hybrid pigs available in this country, not to mention the 'new' breeds recently imported such as the Pietrain, Hampshire and Duroc Jersey. Is it wise to keep pedigree pigs and aim for an 'élite' standard or is it better to buy hybrids—bred with the aid of computerised records—or could one get just as good results if one reciprocally crossed three pure breeds?

Clearly much thought must be given to this problem, but it should be remembered that when looking at the performance of any particular type of pig that it must be related to the conditions in which the pigs were raised. Quite often well-bred pigs perform disappointingly when moved from their home environment to another farm. This is partly explained by the stress factors involved in moving and because the pigs will lack immunity to some diseases in their fresh surroundings. To overcome this problem it is advisable to purchase maiden gilts only and have them served, rather than buying in-pig gilts. This will allow the gilt more time to build up resistance to disease, and she will later pass on this immunity to her pigs via the colostrum (first milk) when suckling her litter.

When choosing your breed also take into consideration the type of production that you are aiming for, the existing or proposed building that will be available, the type of feed you intend to use, and whether the pigs will be kept indoors or outdoors, even if only for part of their time.

Finally, remember that with pigs, like most other commodities, you only get quality if you are prepared to pay a reasonable and fair price. Don't be tempted into buying at 'bargain prices' second-rate stock, which may well prove to be expensive later on.

Six Pig Rations

Introduction

Feeding is one of the most important factors in pig husbandry. Food represents something like 60% of the total cost of breeding and up to 80% with fattening pigs. Great care should be taken to ensure that pigs are fed correctly. Remember that by nature pigs are greedy, wasteful animals, yet, if properly cared for, are capable of rapid growth and are excellent food converters.

The pig has a simple digestive system which cannot deal with large amounts of fibre. This means that pigs must be fed low-fibre, high-energy foods such as cereals, oil cakes, fish, milk and meat by-products, which are relatively expensive.

You must know something of the constituents of foodstuffs, the pig's nutritional requirements, how to compile a balanced ration, and how to mix and feed it, if you are to become proficient in pig husbandry.

In order to measure the amount of energy in any one food or to compare it with another we use either digestible energy (D.E.) or the American method of total digestible nutrients, usually referred to as T.D.N.

Energy

There are many forms of energy, each of which can change from one form to another. For example, green plants convert solar energy into chemical energy. Animals convert the chemical energy in their food into various forms of energy when they digest their meals. Thus energy may be changed into work to allow the animal to walk about, or it may be used to produce heat to keep the body warm or it may be stored in the form of glycogen in the liver or laid down as fat and stored in the body.

53

Scientists are able to calculate the amount of energy in a foodstuff by analysis in the laboratory. A special instrument known as the 'bomb calorimeter' is used. This instrument burns or oxidises a certain amount of food and measures the heat of combustion. The results are recorded in calories and the analysis of the food is given as the *gross energy calorie* (cal).

A calorie is defined as the amount of heat required to raise the temperature of 1 g of water from 14·5 to 15·5°C. For most purposes this is too small an amount and so it is more common to use the kilocalorie (Kcal) which is 1000 calories, or the megacalorie (Mcal) which is 1000 kilocalories.

Digestible energy

Although as we have seen the gross energy of a food can be calculated in the laboratory, not all of this energy will be utilised by the pig and a certain proportion of energy will be excreted in the dung and urine.

A more accurate measurement of energy is, therefore, digestible energy (D.E.), which is defined as *the gross energy (or heat of combustion) of a feed minus the gross energy of the corresponding faeces, expressed as Mcal or Kcal per kilogram of total feed*. This definition is sometimes referred to as apparently digestible energy and means exactly the same.

Sometimes (especially with poultry rations) you will see the term *metabolisable energy* which is gross energy less the gross energy of faeces and *urine*, but, as the amount of energy in urine is very small, and somewhat difficult to obtain, most nutritionalists use the system of digestible energy to express the value of feed.

Total digestible nutrients is the method most widely used for measuring the energy value of foods for pigs. It is calculated as % digestible protein + % digestible fibre + % digestible carbohydrates + (% digestible fat × 2·25).

Nutritive ratio is a term closely related to T.D.N. It means the ratio of energy to digestible protein in a food or ration. For example, wheat has a T.D.N. of 75 and digestible crude protein of 10, the nutritive ratio therefore is 7·5:1.

Constituents of food

Food consists of six main constituents, namely carbohydrates, fibre, fats, protein, vitamins, minerals and water.

Carbohydrates consist of the elements carbon, hydrogen and oxygen, which combine together in various proportions to make simple sugars, starch and fibre. They are utilised by the animal to provide energy to maintain the normal body functions, and keep the animal warm and active. Surplus carbohydrates are stored in the liver, in the form of glycogen, or converted into fat and stored in the body.

Fats and oil may be considered as concentrated carbohydrates. One kilogram of fat has 2·3 times as much energy as one kilogram of pure starch.

Fibre is often referred to as cell-wall constituents. It is a rather complex carbohydrate, the main constituents being cellulose and lignin. In the ruminant, fibre is broken down into simple sugars by the action of the bacteria during digestion, but in animals with a simple stomach, like the pig, very little fibre is digested. Young pigs, especially, should not be fed rations with a high percentage of fibre or their growth rate will be severely retarded. Adult pigs can manage some fibre, but this should not exceed 6% in a ration.

Protein consists of the elements carbon, hydrogen, oxygen, nitrogen and sulphur, which combine together to form amino-acids and amides which then form proteins. Proteins are necessary to build and repair muscle tissue, to promote body growth, and to make milk in lactating sows. There are twenty-two known amino-acids, ten of which are considered essential for normal growth. These are referred to as having 'biological value'. The essential amino-acids are lysine, methionine, tryptophane, histidine, arginine, valine, leucine, isoleucine, phenylanaline and threonine, the most important of these being lysine, tryptophane and methionine.

Experiments at Cambridge and in Australia have shown that the value of a foodstuff depends upon the quantity of essential amino-acids it contains and not on its percentage crude protein. For example, proteins of animal origin, such as fish meal or skimmed milk, have a higher biological value than would the same quantity of vegetable proteins such as groundnut cake or linseed meal. Bellis has shown in feeding trials that adding synthetic lysine and methionine to a ration containing vegetable protein will give as good results as when 7% fish meal is included.

Students are often confused with the various ways in which scientists analyse protein foods, for example crude protein, digestible crude protein, digestible true protein or protein equivalent. It is not necessary here to define each term, but suffice to say that each is a measure of the protein content. The important thing is to remember that when comparing one food with another you should use the same measure. For example, the analysis of white fish meal may be stated as any of the following:

Crude protein	61	Protein equivalent	53
True protein	51	Digestible crude protein	55

Mineral requirements

Under natural conditions pigs consume large amounts of minerals by eating green foods and 'rooting' in the soil. With such an environment, mineral deficiencies are most unlikely to occur, but once the pig is housed indoors, and kept on concrete, careful consideration must be given to the mineral requirements.

For convenience minerals are divided into two main groups— major elements which are required in relatively large quantities and include calcium, magnesium, phosphorus, sodium, potassium, chlorine, iron and sulphur and minor or trace elements which are only required in minute quantities; these are, copper, iodine, manganese, and zinc. There are, of course, many other minerals, but the ones mentioned are those most likely to affect pigs.

The pig is one of the fastest-growing animals, which means that it requires adequate minerals, particularly calcium and phosphorus, for building a strong skeleton. Scientists have stated the exact amounts of each mineral required at the various stages of growth, but it is probably better for the farmer to purchase reputable proprietary mixtures, and add these to their home-mixed rations according to the manufacturer's instructions.

56

Major elements

Calcium

Calcium is required by all types of livestock for the production of bone and assisting in the formation of other tissues. It is important that pregnant and suckling sows and growing pigs have adequate supplies in order that strong bones are built.

A deficiency of calcium leads to bone disorders and in extreme cases rickets may occur. In breeding sows a deficiency will upset the milk supply and with in-pig sows there may be a number of piglets born dead.

Phosphorus

Phosphorus is closely linked with the element calcium in the formation of bones. It is also important in the formation of body cells and in the production of ovum and sperms in the reproductive organs, and in the metabolism of carbohydrates and fats. Cereal grains contain adequate amounts of phosphorus and so cereal-fed pigs are unlikely to suffer from a deficiency.

Sodium, potassium and chlorine

These three minerals are found as salt in the body fluids, sweat, and saliva; and chlorine is needed in the manufacture of hydrochloric acid which is present in the digestive juices.

Vegetable foods are usually rich in potassium, and animal foods, for example fish meal, will supply sodium and chlorine. Where pigs are fed on vegetable foods only it is necessary to include a mineral mixture containing common salt.

Pigs are susceptible to excesses of salt, which can cause salt poisoning. This is unlikely, however, if properly balanced rations are fed and adequate supplies of clean drinking water made available.

Iron

Iron is an essential constituent of blood, and forms a part of haemoglobin which is responsible for carrying oxygen around the body. A deficiency of iron causes anaemia, which can be troublesome in small pigs (see piglet anaemia) when housed indoors. Pigs that have access to the soil are unlikely to be affected.

Minor or trace elements

Trace elements are required in minute quantities, but nevertheless are important in the pig's nutrition. The trace elements are copper, manganese, zinc, cobalt and iodine.

Copper

Copper is associated with iron in the formation of blood and though serious deficiencies can occur in sheep (causing swayback) and cattle (causing scours) there is usually no problem with pigs. However, we do know that feeding supplementary copper to pigs will improve both growth rate and food conversion (see page 78).

Copper toxicity

It is well known that copper salts given in excess to farm animals are toxic. Copper when fed in excess of the pig's nutritional requirements will gradually accumulate in the body tissues, especially the liver, and may eventually cause the death of the animal. For this reason copper supplements should not be fed to breeding stock.

Zinc

A deficiency of zinc leads to a condition in pigs called parakeratosis. The symptoms are subnormal growth, poor food efficiency and skin lesions. There is reddening of the skin, particularly over the belly area, and this is followed by eruptions which develop into scabs. The disease is most common amongst young intensively housed pigs fed *ad lib* on dry meal. The trouble can easily be prevented by adding 150–200 grams of zinc carbonate per tonne to the diet.

Iodine

Iodine is necessary for the thyroid gland, which produces a substance called thyroxine. The thyroid gland is found in the neck and if a deficiency of iodine occurs, the gland swells and the condition is known as 'goitre'.

Thyroxine is used to control the normal body functions, and a deficiency leads to reduced fertility, and in-pig sows may produce dead or hairless pigs. Most foods contain traces of iodine and particularly those of marine origin, for example fish meal and seaweed. Where fish meal is included in a ration iodine deficiency is unlikely to occur.

Manganese

Manganese is necessary in minute amounts for normal reproduction and bone formation. A deficiency can cause lameness in pigs, but this is comparatively rare. Bran and wheat offals provide good sources of this mineral.

Vitamins are complex substances, which have no feeding value in themselves, but greatly help nutrients to function. There are fifteen known vitamins, some of which we term essential for pigs; others may be manufactured by the animals themselves. The essential vitamins for pigs are:

Vitamin A which is extremely important in pigs of all ages. A deficiency in young pigs will lead to poor growth, general unthriftiness, and may affect fertility and milking ability in breeding sows. Vitamin A is found in cod-liver oil, and may be manufactured by the pigs if they are fed green food containing carotene, such as dried grass. Synthetic vitamin A is available and may be added to the ration in a vitamin pre-mix.

Vitamin D is necessary for the proper formation of bones. A deficiency leads to rickets. Growing pigs and pregnant sows, therefore, must receive an adequate amount of this vitamin. Fortunately, vitamin D is manufactured by the pig, provided its skin is exposed to direct sunlight. Therefore, breeding stock and growers kept in open yards or at pasture are unlikely to need supplementary vitamin D. Pigs that are kept indoors, especially during the winter months, must receive vitamin D in their diet. Synthetic vitamin D is available.

Vitamin E is known as the anti-sterility vitamin, since, when it was first discovered, it was found to be essential for the reproduction of rats. However, experiments have shown that with cattle and sheep vitamin E is not concerned with reproduction. Pigs may require it for normal reproduction, but as the vitamin is found in cereal grains, there is no need to feed a supplementary source.

Vitamin B group consists of several vitamins, although pigs seldom suffer from a lack of them. They include vitamin B1 or thiamine, which is found in most foods, and vitamin B2 or riboflavin, found in meat, fish and oil meals.

Vitamin B12, sometimes called the animal protein factor, is important to pregnant sows. A deficiency will lead to low birth weights and high mortality in baby pigs. Rations that contain fish meal will have adequate vitamin B12 for the sow's requirements.

Vitamin C, or ascorbic acid, is synthesised in growers and adults, but there is evidence that baby pigs may need supplementary vitamin C if they do not receive adequate sow's milk.

Reference

The value of the protein in pig food, *B.O.C.M. Pig Information Bulletin,* August (1967).

Seven Common Pig Foods

This chapter briefly describes the more common foodstuffs used in formulating pig rations. For convenience the foods are grouped into protein foods and carbohydrate foods. At the end of the chapter analysis tables are shown. It is most important that the reader should realise that all foods vary in quality and that no figures are 'standard' for a particular food. One should, therefore, regard the descriptions and analysis as being 'typical' of the broad information that is generally required for the purpose of compiling rations.

Protein foods of animal origin

Fish meals

Fish meal is probably the most important protein food used in pig rations. It is manufactured from waste fish and fish offals—head, skeleton and tail. The best-quality fish meal is made from white fish, for example cod, haddock and plaice. White fish meal is guaranteed not to contain more than 6% oil and 4% salt, and at these levels should not cause any taints in the bacon.

Second-quality fish meal made from fish with a high oil content may affect the carcass by causing a taint if fed in excessive amounts. This, however, is most unlikely today because fish meal is expensive to buy and usually restricted in the rations for growing pigs, and omitted in the final fattening mixture.

Herring meal which tends to have a high oil content may safely be fed to breeding stock, if fed to pigs intended for bacon production; remove it from the ration once the pigs reach 50 kg liveweight.

White fish meal may be included (if cost permits) at up to 15% of total for creep rations and around 10% for breeding stock and growers. Besides having a biological value, fish meal also supplies the important minerals calcium, phosphorus and chlorine.

Meat-and-bone meal

Although somewhat variable in quality and supply, meat-and-bone meal is an extremely valuable protein food. It is manufactured from the waste products of slaughter houses and manufacturers of tinned meats and meat products. Usually meat-and-bone meal is sold on analysis of around 40–50% crude protein and is guaranteed not to contain more than 4% oil. It may safely be included at up to 10% of the ration.

Dried blood

Rarely obtainable today, dried blood is a by-product of the slaughter house. It is first dried at very high temperatures in order to overcome the risk of disease. Dried blood has a very high feeding value, and is especially suitable for growing pigs, but should not be fed in excess or scouring may occur: $2\frac{1}{2}$–5% of the ration would appear to be most suitable.

Separated milk

Skimmed or separated milk is a by-product of butter making. It contains all the original solids of whole milk except for the reduced butterfat content. It is a very valuable foodstuff, especially for suckling sows and youngsters. It is rich in animal protein (35% of dry matter) and contains a high lysine content. It also has a high lactose (milk sugar) content and so it can replace part of the cereals in the ration.

The Milk Marketing Board claims that feeding skim milk will improve the carcass grading, and that whenever checks have been made at bacon factories to compare non-skim with skim-fed pigs the results have always shown a higher proportion of top grades for the skim-fed pigs.

Where regular supplies are available skim milk may be fed with finely ground barley meal—allow 2 litres of skim per pig at weaning and gradually increase to 4–5 litres at bacon weight. Feed as much barley meal as the pig will clear up in twenty minutes, twice daily. Suckling sows may receive 10–15 litres per day.

Skim may be fed fresh (sweet) or sour (preserved). Where the supply has to be kept for more than one or two days it is best to be preserved by adding 0·1% formaldehyde.

Twenty-five litres of skimmed milk is roughly equivalent in feeding value to 1 kg of fish meal and 2 kg of barley meal.

Dried separated (skim) milk

Due mainly to the high cost of drying, dried skim milk tends to be rather expensive.

Dried separated milk comprises about 35% protein, and contains minerals, plus a little fat. It also contains some vitamins, particularly riboflavine. It is used mainly for making milk replacement meals for rearing baby piglets and is sometimes included in creep-feed rations.

Protein foods of vegetable origin

Vegetable protein foods are less expensive than those of animal origin, and are widely used as partial or complete replacements. Although the amino-acid content of these foods are not so well suited to the pigs needs as that of animal proteins, satisfactory results can be obtained by substituting part of the ration in the growing stage and complete replacement in the finishing stage.

Soya-bean meal

Soya-bean meal is the residue after the oil has been removed from imported soya beans. It has a crude protein content of around 38% which is slightly lower than groundnut meal at 41%. However, soya bean is particularly rich in lysine, an important amino-acid and this makes it the most popular source of vegetable protein for pig rations. In appearance it is a pale yellow colour and has a 'gritty' texture. It is low in fibre and this can have a laxative effect on stock if fed in excess. It is low in calcium and so a mineral mixture should be added to the ration when it is used to replace fish meal. Normally soya-bean meal is included at 5% with 5% fish meal in growing rations and at 10% as the sole supply of protein in finishing rations.

Decorticated groundnut meal

Decorticated groundnut meal is the residue from imported groundnuts after the oil and outer husk has been removed. It is probably the most widely used protein food for cattle, but is less suited to pigs than soya bean because of its lower lysine content. However, if soya bean is scarce or too expensive, groundnut meal may be advised. Like soya bean it is low in calcium, and a mineral mixture should be added to the ration when it is used to replace fish meal.

Dried yeast

Dried yeast is a by-product of the brewing and distilling industry. It has a crude protein of around 40% and is particularly valuable for growing pigs because of its high vitamin B content. For this reason it is often recommended for feeding with swill, which is low in vitamin B.

Yeast is also available in a liquid form, in which case six litres of liquid yeast may replace 1 kg of fish meal.

Peas and beans

These two foods are both home grown and have a crude protein content of between 19 and 20%. Beans are quite widely grown for stockfeed but peas are usually those which are unsuitable for human consumption.

They may be included in fattening rations at up to 20%, but care may be needed with beans as they tend to have costive effect. Both peas and beans produce carcasses with a white firm fat.

Carbohydrate foods

Barley meal

Barley meal is without doubt the most important carbohydrate food and usually consists of 50% or more of the pig's diet. It is,

64

of course, a home-grown cereal, although large quantities are imported as well. In general it has a low oil and fibre content, and is rich in starch. It is important to remember that feeding values do vary with the quality of grain, and the farmer should look for plump samples of barley with a thin skin that shows good colour. By cutting the grain it is possible to observe the starch content—a full white grain indicates a good sample.

Always avoid buying shrivelled, thick-skinned grain with a 'weathered' appearance.

Ground wheat

For a long time wheat was thought to be unsuitable for pig feeding, but in recent years wheat has been used in ever-increasing amounts, especially for fattening stock. Provided that it has been well harvested, wheat may be included at up to 50% of the carbohydrate foods, and some farmers feed more. Care should be taken with newly harvested wheat with a high moisture content, as this may lead to scouring if fed in excess. Wheat requires grinding coarsely, as finely ground wheat meal is likely to be pasty and cause digestive upsets.

When using a hammer mill a 4 mm screen is recommended. Wheat is rich in vitamin B and is higher in protein than barley.

Maize meal

Maize meal has less fibre but more oil than barley and for this reason it has a higher energy value. If fed in excess to fattening pigs the high oil content may lead to soft oily fat in the carcass. Maize is usually restricted to around 25% of the ration and should never exceed 40% or nutritional upsets and poor-quality carcasses may occur. Yellow maize contains a substance which the pig converts into vitamin A.

Flaked maize

Flaked maize is produced by cooking whole maize and then passing it through rollers. The process produces extremely palatable and highly nutritious flakes. This makes flaked maize a valuable food for young pigs and sick animals where palatability is important.

It is often included as a 'tracer' food when mixing two rations on a farm, thus, for example, the sow and weaner ration will have 'yellow flakes' and the fattening ration will be identified by the absence of flaked maize.

Oats

Oats are of limited value in pig feeding due mainly to their low energy value and high fibre content when compared with either barley or wheat. They may be fed to adult breeding stock or heavy feeding pigs. Some farmers include oats in the fattening ration of bacon pigs fed on an *ad lib* basis, in order to control the energy intake and improve carcass grading. Oats should be ground finely, and restricted to about 10% of the ration.

Wheat offals

Wheat offals are the residue left after milling. The extraction rate of flour obtained from wheat varies, but generally is around 70%; this leaves 30% offals, which are usually in the ratio of two parts middlings (weatings) to one part bran (the outer coat).

Bran

There are two types of bran on the market—ordinary bran and broad bran. Bran has many qualities other than what its analysis would suggest. For example, bran may be fed as a 'bran mash' to sick animals or as a laxative to sows just before farrowing. On the other hand, bran may be fed dry to scouring pigs to act as a 'binder'.

In general, bran is not included in the normal rations because of its bulk in relation to feed value and its high fibre content.

Wheat feed or middlings

Wheat feed or middlings has many brand names and is often referred to as weatings, superfine weatings, pollards, sharps, or thirds. It can be included in any pig ration, as it improves the texture and palatability and also acts as a 'safe' food in preventing digestive disorders. The only disadvantage of wheat feed is that it frequently costs more to purchase than whole wheat! and, of course, its nutritional value is considerably less.

Where wheat feed is obtainable at a realistic price it may be included with advantage at up to 30–40% of a ration for growing stock and around 25% for fattening pigs.

Whey

Whey is a by-product of cheese making and is essentially a carbohydrate food, twelve litres of whey being approximately equal in feeding value to 1 kg of barley meal. It is a highly digestible feed, mildly laxative and is usually fed 'sour'. It is an ideal food for pork or heavy-hog production, especially where it can be included in a 'pipeline' feeding system. Whey can also be obtained in a dried form, and as whey paste which has a moisture content of about 34%.

Potatoes

Stockfeed potatoes can be offered as a suitable replacement for barley meal—4 kg of potatoes are equal to 1 kg of barley. They are best fed after boiling or steaming for twenty minutes, or if pulped by machine—both operations make the potatoes more palatable and improve digestibility. It is unwise to feed potatoes to growing pigs until they are eating $1\frac{1}{2}$ kg of meal per day.

As a guide to the purchase price, one can reckon on potatoes being economic if purchased at one-sixth the cost of barley meal. This will allow for the extra labour in handling and preparation.

Swill

There are innumerable types of swill, ranging from predominantly waste cabbage and lettuce leaves to high-grade swill containing a large proportion of meat and fish waste in addition to potatoes and other vegetables, the main detraction against swill thus being its variable quality, the expense of handling, processing and smell! By order of the *Diseases of Animals (Waste Foods) Order 1957*, all swill must be boiled or steamed for at least one hour (boiling means kept at a temperature of not less than 100°C) before it is fed to pigs.

Swill should be fed in conjunction with meal, and as with feeding potatoes it is best if the pigs are first eating 1½ kg of meal per day before the cooked swill is offered. Generally one can reckon on 3 to 4 kg of 'best' swill and upwards of 6 kg of 'second quality' swill replacing 1 kg of barley meal.

Eight Balancing Rations

There are two main methods of formulating pig rations in general use today: the digestible energy method and the T.D.N. system. Scientists and feedingstuff manufacturers use the digestible energy system, which is the more accurate of the two, but most farmers still adopt the T.D.N. method, which is simpler and easy to use when either buying foodstuffs or compiling rations on the farm. This chapter will first explain the digestible energy system and then give examples of rations using T.D.N.

All pig meals may consist of five main ingredients, namely cereals to provide energy, animal and vegetable protein foods, minerals and vitamins. In some special rations 'additives' may also be included. It is our task to see that the ration contains the correct proportion of each ingredient in order to satisfy the pigs' requirements.

The Agricultural Research Council, in their publication *The Nutrient Requirements of Farm Livestock, No. 3 Pigs*, give the following standards for growing and fattening pigs.

Table 6. Protein requirements for growing and finishing pigs

	Growing pigs (weaning to about 50 kg)	Fattening pigs (50–90 kg)
Crude Protein	16·10–17·4%	13·00–14·40%
Lysine	0·78– 0·83%	0·61– 0·65%
Methionine and cystine	0·52– 0·61%	
Tryptophan	0·13– 0·16%	
Threonise	0·44– 0·52%	
Isoleucine	0·65%	

Table 7. *Compositional and nutritive aspects of feeds*
(Figures expressed on an air-dry basis)

	Barley	Wheat	Maize	Oats	Wheat offals	White fish meal	Soya-bean meal
Digestible energy (D.E.) (Mcal/kg)	3·12	3·41	3·41	2·62	2·9	2·62	3·21
Crude protein%	10·0	12·0	9·5	10·5	15·5	63·0	45·0
Total lysine%	0·40	0·35	0·27	0·07	0·70	4·5	2·50
Total methionine and cystine%	0·50	0·66	0·39	0·36	0·60	2·45	1·33
Total tryptophan%	0·15	0·10	0·05	0·12	0·02	0·66	0·51
Total threonine%	0·43	0·36	0·48	0·37	0·50	2·54	1·82
Total isoleucine%	0·54	0·53	0·38	0·46	0·55	2·43	2·07
Calcium (Ca)%	0·05	0·04	0·01	0·10	0·09	7·1	0·21
Phosphorus (P)%	0·37	0·38	0·36	0·35	0·66	3·9	0·90
Chlorides as NaCl%	0·19	0·13	0·12	0·12	0·12	1·6	0·05
Zinc p.p.m.	28	20	20	32	106	104	61
Manganese p.p.m.	12	21	3	44	84	6	31

From the table you will notice that young pigs require 16–17% crude protein and approximately 0·80% lysine. Whilst the other amino-acids are important, their respective amounts are so small that we may safely assume that sufficient will be available if we include some animal protein food in the ration. The fattening pigs need for protein is slightly less at 13–14·4% and lysine at 0·61–0·65.

Energy requirements

The A.R.C. recommend that pig meals should contain about 3·0 Mcal/kg. In the United Kingdom barley meal is the main cereal used and frequently accounts for over 60% of the ration. In Table 7 you will see that barley has an energy value of 3·12 Mcal/kg, whilst wheat and maize is slightly higher and wheat offals are lower in energy than barley. In practice we find that by including a high proportion of barley meal the end product will have a digestible energy of approximately 3·0 Mcal/kg.

The following worked example should illustrate this point:

Example growing ration

%				Digestible Energy
60 Barley meal	=	6×3.12	=	18.72
20 Wheat offals	=	$2 \times 2·90$	=	5·80
10 Maize	=	$1 \times 3·41$	=	3·41
10 Fish meal	=	$1 \times 2·62$	=	2·62
		10		30·55

$$\frac{30·55}{10} = 3·05 \text{ digestible energy}$$

Model rations

We can now proceed to formulate diets, omitting for the present minerals and vitamins, which are discussed later. A simple model ration based on barley meal, soya bean, and fish meal for (A) growing pigs and (B) fattening pigs suggested by *Dr. A. Eden* is as follows:

	Model rations		
A		**B**	
Growers		Fatteners	

	%		%
Barley meal	85·0	Barley meal	92·5
White fish	7·5	White fish meal	2·5
Soya bean	7·5	Soya-bean meal	5·0

On the basis of Tables 4 and 5 these rations
have the following analysis

	A	B
	%	%
Crude protein	16·6	13·1
Lysine	0·86	0·61
Methionine and cystine	0·71	0·57
Tryptophan	0·22	0·18
Threonine	0·69	0·55
Isoleucine	0·80	0·66
Digestible Energy	3·08 Mcal D.E./kg	3·08 Mcal D.E./kg

Source: *Nutrient Requirements of Pigs—Dr. A. Eden.*

These rations conform reasonably well to the estimated
requirements of crude protein, lysine and digestible energy.
The inclusion of some wheat offals—say 10–15%—would
raise slightly the protein and amino-acid content. If maize were
included it would raise slightly the digestible energy, but lower
the protein.

We must always remember, however, that foods vary in
analysis, according to their 'quality' and storage condition. To
offset this variation, the inclusion of several ingredients in a
ration is recommended.

Feeding scales

Having formulated our rations, the next step is to calculate the
daily amount to be fed to each pig in accordance with its liveweight and expected growth rate. The National Pig Progeny
Testing Stations standards for pigs fed twice daily on wetted
meal, and where the complete amount is eaten within about
twenty minutes, is given below in Table 8. You will see, for
example, that a pig of 20 kg liveweight is expected to put on

about 500 g liveweight daily. To do this it will need 3·00 Mcal digestible energy, which is provided in 1 kg of meal. At 50 kg liveweight the pig will need 6·60 Mcal digestible energy, this is provided in 2·20 kg of meal, and so on.

Table 8. *Average daily feed intakes of pigs fed twice daily which follow the average growth curve*

Liveweight	For the following week Expected daily liveweight gain	Daily feed intake		
(kg)	(g)	(kg diet)	(kg dry matter)	(Mcal D.E.)
20	500	1·00	0·87	3·00
25	550	1·20	1·04	3·60
30	625	1·45	1·26	4·35
35	690	1·65	1·44	4·95
40	750	1·85	1·61	5·55
45	775	2·05	1·78	6·15
50	790	2·20	1·91	6·60
55	790	2·35	2·04	7·05
60	790	2·50	2·17	7·50
65	790	2·65	2·31	7·95
70	790	2·75	2·39	8·25
75	790	2·85	2·48	8·55
80	790	3·00	2·61	9·00
85	790	3·10	2·70	9·30
90	790	3·20	2·78	9·60

Source: *The Nutrient Requirements of Farm Livestock, No. 3 Pigs* Agricultural Research Council (1967).

High levels of feed intake such as these encourage rapid growth rate, but may also lead to increased fat content in the carcass during the later stages of fattening. To this end many farmers restrict the meal allowance from around 50 kg liveweight to slaughter weight, feeding not more than 2·7 kg per day.

Total digestible nutrients and digestible crude protein

The second method of balancing rations is to use tables of T.D.N. and digestible crude protein (D.C.P.) to formulate suitable meals for the various classes of pigs. Table 9 gives the analysis of the more popular pig foods and Table 10 suggests suitable standards.

Table 9. Nutritive values of common pig foods

	D.M.	T.D.N.	D.C.P.	Fibre
Cereals				
Barley meal	86	71	8·6	4·8
Maize meal	87	78	7·5	2·0
Flaked maize	89	86	10·0	1·5
Oats (farm ground)	87	62	9·6	10·0
Sorghum (ground)	87	79	8·2	2·3
Wheat	87	75	10·0	5·0
Oil cakes and meals				
Ground nut meal (decorticated)	88	73	5·0	5·2
Palm kernel meal (extracted)	90	56	11·4	16·0
Soya-bean meal (extracted)	89	72	38·5	5·0
Legumes				
Bean meal	86·8	66·4	22·2	9·6
Feedingstuffs of animal origin				
Fish meal—white	87·5	63	59·1	—
Meat meal	92·1	84·7	61·9	—
Meat meal (extracted)	93	68	58·6	—
Whole meat	94·4	85·6	76·9	—
By-products				
Miller's offals	87	51·4	9·0	10·3
Fine wheat feed	87	63·4	12·7	7·5
Coarse bran	87	48·2	10·3	10·3
Swill				
Urban swill (processed)	31·9	28·1	2·9	1·8
Military-camp swill	25	22·4	2·6	1·5
Potatoes, boiled	No published figures available for T.D.N., but it would be reasonable to assume T.D.N. 18 as 4 kg of cooked potatoes will replace 1 kg of barley meal.			

N.B. [1] The above figures are taken from H.M.S.O. Bulletin 48

[2] These figures should be taken as a guide only, because individual samples of foods will vary in composition and digestibility

Again we make the assumption that foods with a high protein value will also be richer in amino-acids, and you will note that young pigs must be fed on low-fibre diets.

A well-tried method is to find the nutritive ratio, N.R., needed by the type of pigs you wish to feed (see Table 8) and then prepare a 1-tonne mixture that conforms. The arithmetic will be much easier if we use 10–100 kg mixture, as each part will represent 10% of the ration.

Example 1. Compile a weaner ration

If you look at Table 10 you will find that weaner baby pigs require a mixture with N.R. of 5:1

%	Parts or 100 kg		T.D.N.	D.C.P.	T.D.N.	D.C.P.
10 Fish meal	=	1 × 63	59	=	63	59
20 Weatings	=	2 × 63	12	=	126	24
20 Flaked maize	=	2 × 86	12	=	162	20
50 Barley	=	5 × 71	8·6	=	355	43
100		1 tonne			706	146

The 1-tonne mixture has 706 units T.D.N. and 146 units D.C.P. which means that the analysis of 100 kg would be 70 T.D.N. 14·6 D.C.P. with a N.R. of 5:1.

Remarks: This is an ideal ration for young pigs. The fish meal has a high biological value and provides minerals, particularly calcium and phosphorus. The flaked maize is high in energy and provides a precursor of vitamin A.

Table 10. Nutritive requirements

Ration	Age	T.D.N.	D.C.P.	N.R.	Maximum fibre %
Creep feed	2– 8 weeks	70–72	16	4·5:1	3·0
Weaner	8–16 weeks	70–72	14	5:1	3·5
Breeding stock					6·0
Growers	16–20 weeks	70–72	13	6:1	4·0
Baconers	20–24 weeks	70–72	12	6:1	5·0
Heavy hogs	16–30 weeks	70–72	10	7:1	5·0
Dry sows	Adults	70–72	7	10:1	5·0

Weatings will 'lighten' the ration. Barley meal completes the ration; it is relatively cheap and forms half of the mixture.

Example 2. Grower ration

%	Parts or 100 kg	T.D.N.	D.C.P.	T.D.N.	D.C.P.
5 White fish meal =	0·5 ×	63	59 =	31	30
5 Soya bean =	0·5 ×	72	38·5 =	36	19
10 Weatings =	1 ×	63	12 =	63	12
20 Wheat =	2 ×	75	10 =	150	20
60 Barley meal =	6 ×	71	8·6 =	426	51·6
100	1 tonne			706	132·6

Remarks: This ration has 70 T.D.N. and 13·2 D.C.P. with a N.R. of $5\frac{1}{2}$:1 which is suitable for growing pigs.

Example 3. Fattening ration

%	Parts or 100 kg	T.D.N.	D.C.P.	T.D.N.	D.C.P.
5 Soya bean =	0·5 ×	72	38·5 =	36	19·25
15 Weatings =	1·5 ×	63	12 =	94·5	18
20 Maize meal =	2 ×	78	7·5 =	156	15
60 Barley meal =	6 =	71	8·6 =	426	51·6
100	1 tonne			712·5	103·85

Remarks: This ration has 71 T.D.N. and 10 D.C.P. with N.R. 7:1. Note how the fish meal has been replaced by soya-bean meal to lower the cost. A proprietary mineral should be added to this ration.

The examples shown here should be taken as a guide only to formulating rations. It is not claimed that these rations will suit all pigs or all conditions of management. But as you gain experience you will be able to modify and change rations according to the strain of pig you keep and the type of production you are aiming for.

Remember, also, to cost your ration and to calculate the final cost of food per kilogram liveweight gain.

Buying raw materials

When purchasing 'straight' foods such as barley meal, flaked maize, or maize meal it is worth while calculating the cost per unit of T.D.N., for example, if barley meal cost £30 per tonne, maize meal £34 per tonne, wheat £33 tonne and wheat offals £30 tonne. Which would be the best buy?

$$\text{Barley} = \frac{£30}{71 \text{ T.D.N.}} = 40\tfrac{1}{2}\text{p per unit}$$

$$\text{Maize meal} = \frac{£34}{78 \text{ T.D.N.}} = 46\text{p per unit}$$

$$\text{Wheat} = \frac{£33}{75 \text{ T.D.N.}} = 44\text{p per unit}$$

$$\text{Wheat feed} = \frac{£30}{51 \text{ T.D.N.}} = 59\text{p per unit}$$

Clearly from this list barley is the least expensive and wheat feed the most expensive.

Protein vitamin mineral premixes are available from commercial feedingstuff manufacturers. They consist of carefully blended proteins with the correct amount of minerals and synthetic vitamins added. The premix is mixed with home-grown cereals to produce a balanced ration.

Milling: Barley grain should be finely ground. If using a hammer mill, a 3 mm sieve should be used.

Wheat should be ground coarsely, kibbled, or rolled. Finely ground wheat will change into dough in the pig's stomach and cause digestive upsets.

Oats have a high fibre content and should be ground as finely as possible.

Mixing: Small quantities of food may be mixed by hand, especially if a premix is used. The ground cereals should be heaped on a clean floor and the premix spread on top. Using a lightweight shovel, turn the meal at least twice, preferably three times, to give a thorough mix.

Where large quantities of food are to be mixed, or where feed additives such as copper or antibiotics are included, a mechanical mixer is recommended.

Feed additives

Copper

Some years ago *Dr. Braude* established that young pigs prefer diets which include a small proportion of copper, or copper sulphate, and that their growth rate was improved. In later trails Dr. Braude reported that feeding 1 kg of copper sulphate per tonne (250 parts per million) improved food conversion by 9·7%, and liveweight gain by 7·9% over the controls.

Copper should only be fed to growing pigs intended for slaughter. Breeding stock should not be allowed copper, because surplus amounts are stored in the liver and could result in copper poisoning if fed continuously for a long period. It is important that the copper is very thoroughly mixed.

Antibiotics and arsenicals

Although generally described as growth promoters, antibiotics and arsenical food additives are really disease inhibitors. The action of these drugs is to control bacteria in the gut, which is living on the pig's food, and to inhibit the growth of disease-forming bacteria. Pigs vary considerably in their response to feed additives. Some (usually the runts) make spectacular live-weight gains, while others make little response.

The use of feed additives is now carefully controlled by the Government, following the Swann Report in 1970. Certain antibiotics, for example penicillin, chlortetracycline and oxytetracycline, are only available on a veterinary surgeon's prescription. Others are freely available, such as zinc bacitacin, flavomycin and virginiamycin.

The inclusion of a feed additive is well worth while where pigs are subjected to periods of stress, such as being mixed in large weaner pools, or when housed temporarily in poor conditions, but in the long run it is wiser to avoid continual use if possible, and then to use the prescribed drug should an outbreak of disease occur. Prolonged use of any drug is liable to lead to a resistance to it by the disease-forming bacteria.

Summary

1 Carbohydrates provide—energy, warmth, maintain body functions. Surplus builds fat.
2 Proteins—build body tissue, stimulate growth and milk production, affects the proportion of lean meat (muscle) in the carcass.
3 Vitamins and minerals—essential for building strong skeleton and maintaining growth.
4 Rations must be balanced with the correct ratio of protein to carbohydrates, according to the pig's nutritional requirements.
5 Rations should be carefully mixed, and fed in fresh condition.
6 Copper sulphate may be added at not more than 1 kg per tonne in growers' rations as a growth stimulant.

Nine Pig Feeding

Feeding systems

There are innumerable ways of feeding pigs, whether they are for pork, bacon or heavy hogs. The farmer, in choosing a system, should consider the type of meat production he is aiming at, the breed or cross-breed of pig to be kept, the buildings available, the foods available and not least the aptitude and ability of his pigman. Some of the more popular pig-feeding sytems are now described:

Danish wet feeding

Dry feeding from hoppers

Floor feeding with cubes or meal

Wet feeding with pipeline

The Danish system is widely adopted by farmers who wish to produce quality bacon pigs. Finely ground meal is first placed in the trough and then damped with a little water. This encourages the pig to eat its meal allowance first; it is given as much water as it requires later. The meal allowance is carefully rationed according to the pigs' liveweight. In the early stages the ration allows maximum growth, but after the pig is 70 kg the allowance is restricted to prevent the pig becoming over fat.

A useful guide is given in Table 11.

The bacon × dual-purpose breeds, for example Large White × Saddleback, should be restricted to 2·5 kg per day from 60 kg liveweight onwards. This will prevent overfatness and improve carcass grading.

A simple way to remember this system is to allow half a kilogram of meal per month of age, up to a maximum of 2·7 kg daily.

Table 11. Daily meal allowance for bacon pigs

Liveweight	Daily meal allowance
(kg)	(kg)
20	1·0
25	1·20
30	1·45
35	1·65
40	1·85
45	2·00
50	2·2
55	2·35
60	2·5
65	2·6
70	2·7
75	2·7
80	2·7
85	2·7
90	2·7

Some pigmen ration their stock by feeding as much meal as the pig will clear up in twenty minutes, twice per day.

Young store pigs should average about 2 kg weekly live-weight gain with a food-conversion ratio of 3 kg of meal per kilogram weight gain. In the latter stages bacon pigs will put on around 3–4 kg liveweight weekly, but the food conversion will probably widen to around 3·5–4·0 kg meal per kilogram gain. Obviously this will depend upon the housing, health and management of the herd.

A modification of the Danish system is to feed the rationed amounts of meal entirely dry, and to provide the pigs with a separate water supply such as an automatic drinker or water bowl.

Dry feeding from hoppers is an ideal system for pork pigs, heavy hogs and baconers up to 45 kg l.w. The system allows the pig constant access to meal, and, in so doing, encourages maximum liveweight gain. The hoppers need filling only every day or so, and this saves the pigman the daily chore of twice feeding. However: a note of caution. The great disadvantage of self-feeding is that it is more difficult to spot the ailing pig, since unlike wet feeding you don't see them come to the trough. It is most important, therefore, that the pigman should walk through the pens each day and rouse any suspicious animal to make sure that all is well. Special attention must also be paid to see that the water supply is always freely available.

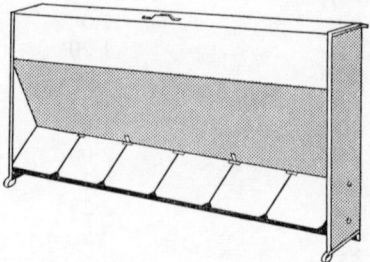

Fig. 19(a).   Single-sided dry-meal pig feeder

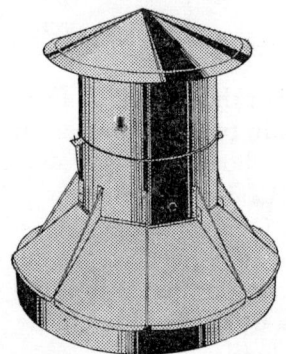

Fig. 19(b).   Round self-feeder

Floor feeding is a fairly recent development in pig husbandry, originating in Ireland. The chief advantage of the system is not in the actual feeding but in the capital saved by not building troughs in new buildings. Also, more pigs can be kept in a pen, since no trough room is required.

Either dry meal or cubes may be used. Feeding consists of simply scattering the food on the floor and allowing the pigs to find it. Water is usually supplied by placing a water bowl in the dunging passage. With this system it is best to build the pens long and narrow with a fairly steep slope in the floor, running towards the dunging area. This will help to keep the pigs from soiling the floor area on which they feed.

Fig. 19(c).　Floor feeding

Fig. 19(d).　Pipeline system of wet feeding

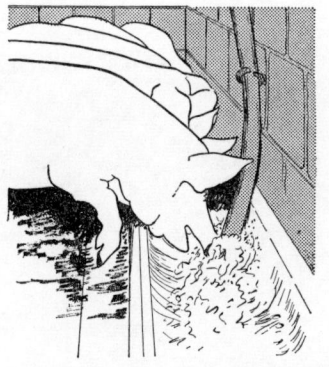

Fig. 19(e).　Wet feeding

Wet feeding has the advantage of making the meal more appetising and easier for the pig to digest. The meal may be mixed into a thick slop, either directly in a bucket at feeding time or in a large tank for all the pigs some time before feeding. Pigs prefer to be wet fed, but the system is very demanding on labour and time.

For large units it is now possible to install a pipeline wet-feeding system. This entails a central mixing unit where pre-determined amounts of water and meal are mixed together mechanically. The wet mix is then pumped around to the pens by pipeline, any surplus returns to the mixer. The pigman simply turns a handle to release the food into the trough. The amount fed is regulated by the pigman 'timing' the amount of food released by using a stop-watch. The only disadvantage of pipeline feeding is the risk of disease bacteria entering the pipeline, particularly in the warm weather.

Ten Selection of Breeding Stock —the Sow

Success in pig keeping depends largely upon keeping good mothering sows. The farmer should, therefore, pay great attention to the selection of his breeding stock. The most important factors in a breeding herd are:

Health   Mothering ability
Prolificacy   Constitution
Teat number   Temperament
Carcass quality of the offspring

Health

Good health is of paramount importance in pig keeping, for unthrifty pigs that grow slowly will quickly lose the farmer money.

The best indications of health are appetite, growth and size. Pigs that are alert, come to the trough quickly, and grow well should be the ones that are selected for future breeding.

Prolificacy and mothering ability

Sows should not only produce large litters but must be able to rear them. Some sows produce large litters but are poor milkers, others are clumsy and step on their pigs, or crush them to death by lying on them. When selecting gilts, compare their dam's performance with M.L.C. Pig Records (see Table 12).

Suitable records would be:
Number born alive   12–14
Reared to three weeks   10–12
Litter weight at three weeks   60 kg
Number weaned at six or eight weeks   9–10
Average weight at six weeks   13 kg

85

Average weight at eight weeks   18 kg

There is some evidence to suggest that the heavier a piglet is at birth then the better it will do throughout its life. Some breeders only select pigs weighing over 1·5 kg at birth for future breeding stock. Birth weights of up to 1·8 kg have been recorded.

Table 12. Pig-feed recording service October 1969–September 1970

Efficiency measures (eight-week weaning)	Good	Average	Poor
Number of litters per sow per annum (farrowing interval)	2·0	1·8	1·4
Number of pigs born alive per litter	12	10	8
Number of pigs reared per litter	11	8	6
Total weight of litter at birth (kg)	18	15	11
Average weight of pigs weaned (kg)	18	16	14
Meal equivalent per pig reared to weaning (kg)	82	104	122
Food cost per pig reared to weaning(a)	£3·50	£4·00	£4·50
Food consumption per kg of weaner production (kg)	2·0	2·5	2·9

Source—adapted from *M.L.C. Pig Facts* (1971)

Teat number and placement

Whilst the N.P.B.A. have insisted upon pedigree sows having at least twelve teats before they are registered, most farmers will agree that fourteen teats are desirable. Sows that produce more pigs than they have functional teats rarely make a satisfactory job of suckling their litter with two piglets suckling one quarter.

The teats should be evenly spaced along the underline, with three pairs of drills in front of the navel.

Temperament is most important. Quiet strains of pigs are invaluable, especially when a sow is farrowing. 'Wild', nervous, bad-tempered animals should be sold for slaughter.

Constitution in pigs means the ability to overcome stress. Sows especially must be able to stand up to the strain of producing and suckling two litters or more a year, perhaps in second-rate buildings and with only 'average' standards of management. Sows running out of doors have problems of climate, altitude and soil type to contend with. They are expected to thrive on the minimum amount of food whilst pregnant, and to 'milk off her back' when suckling. They are only allowed three or four days' rest between weaning a litter and starting the next. Some farmers even have the sows served three weeks after farrowing; no doubt in the interests of greater efficiency!

If the sow does not stand the pace then she is sold, and replaced by a young gilt immediately. In other words, she lacks constitution.

Table 13. Average number of litters farrowed per sow per year

Litters farrowed	Per cent of herds
1·4 and under	12
1·5	6
1·6	10
1·7	15
1·8	17
1·9	16
2·0	10
2·1 and over	14
	Total 100

Average litters farrowed   1·8
Number of Herds   724

Source: *M.L.C. Pig Feed Recording Service,* October 1969–September 1970

Carcass quality

Breeding stock are kept to produce bacon and pork pigs that will have good carcass quality. One of the best ways of improving performance is to select gilts whose litter brothers have been slaughtered and given favourable results.

On research farms, gilts are tested with ultrasonic backfat testing apparatus. The machine determines the amount of backfat in the live pig, which gives a good indication of what the carcass would be like if the gilt were slaughtered.

It may well be that in the future portable ultrasonic testing machines may be available to all commercial pig breeders.

Visual assessment

It is usual to select gilts when they are six months old or of

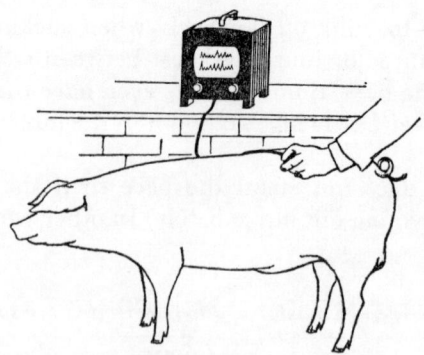

Fig. 20. Ultrasonic testing

bacon weight. Provided the gilt has satisfactory records, she should next be subjected to a visual assessment. It is best to turn several gilts into a yard together in order to study how they walk, and to compare one with another.

A good gilt will be well grown for her age, weighing at least 90 kg at six months.

The head should be light and typical of the breed.

The shoulders should be smooth and neat, fairly wide across the top.

The shoulders should run smoothly into the back, which should be long and slightly arched.

The hams should be wide and deep, with the tail set high.

The legs should be strong and straight, and placed under each corner of the pig. The pig should stand up on well-formed feet, with no sign of weakness in the pastern joints.

The underline should be soft and silky to the touch, and carry fourteen well-developed teats, evenly spaced.

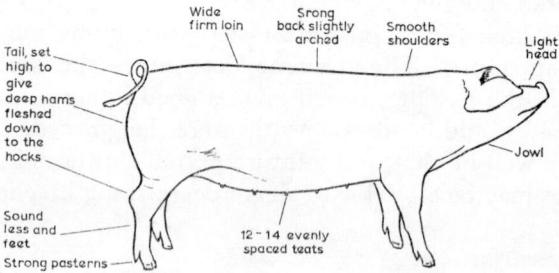

Fig. 21. Points of breeding gilt

Eleven Selection of Breeding Stock —the Boar

The boar is undoubtedly the most important single animal in the breeding herd, and it is no exaggeration to say that his influence can mean the difference between profit and loss when producing bacon pigs. A boar is generally kept for three or four years in a herd, and may during that time sire 2000 or more pigs. We know that food-conversion ratio and carcass quality is influenced by the boar, and this can mean a difference of £1–£2 per pig between good and poor grading. With food conversion, the saving of 0·1 kg of food per kg/l.w.g. will make a difference of approximately 60p per pig fed to bacon, for example.

Boar A's pigs have f.c.r. 3·5 kg meal per kg/l.w.g. This means that 260 kg of meal will be needed to feed a weaner to bacon weight.

Boar B's pigs have f.c.r. 4 kg meal per kg/l.w.g., in which case 300 kg of meal will be required to feed a weaner to bacon weight, or 40 kg more than progeny of boar A's.

Selection and purchase of boars

1 Where possible buy a performance-tested boar, with good records (see page 91 for boar-performance testing).

2 Study your breeding sows carefully and note any weakness, such as poor hams, coarse shoulders or bad legs. If you find anything lacking in conformation, then you can select a boar that is particularly good in the characteristics that you wish to improve in your herd.

3 Milking ability can be improved in a herd by selecting a boar out of a deep-milking sow. We recognise milking ability

by studying the three-week weights of a litter, which should be over 55 kg.

Visual assessment

Never buy a boar on records alone. Always inspect the animal and satisfy yourself that he has sound legs and feet, and appears to have constitution.

The pig should be masculine without undue coarseness; he should be well grown for his age: at least 100 kg at six months.

He should stand firmly on all legs, particularly the hind legs, which must be short, straight and strong. If there is the slightest suspicion of any weakness in the legs or feet, disregard the pig immediately.

Have the boar paraded on concrete and at pasture, to ascertain how he walks. Ask yourself, is this pig capable of getting around muddy fields in winter and serving sixty or more sows per year? A boar must have good legs, walk well and be fit and active to get the sows in pig.

Look at the general conformation, which should be similar to the ideal bacon pig, namely light but masculine head, neat shoulders, long slightly arched back, deep well-flushed hams, and smooth underline with fourteen evenly spaced teats. The underline is important, since a boar will transmit this characteristic to his daughters.

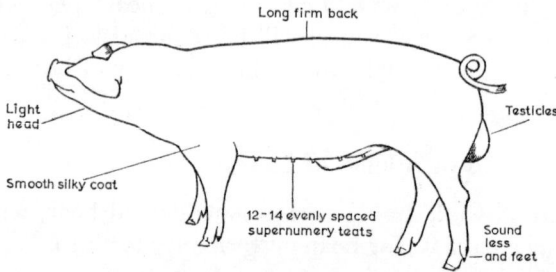

Fig. 22(a).   Points of breeding boar

Summary

The boar has a tremendous influence in a herd. Buy the best pig that you can afford. Inspect the records of the boar's sire and dam and try to see as many of his close relatives as possible. Choose a strong, masculine, well-grown pig that walks well.

Boar-performance testing

This is a system whereby the boar itself is tested before it reaches breeding age. In nucleus herds well-bred young boar pigs may be reared and their performance recorded at one of the Meat and Livestock Commission's Pig Testing Stations. The procedure is that the breeder sends to the station a group comprising of two boars, one castrate, and one gilt—all from the same litter—when they are about ten weeks old. The main objective of boar testing is to assess the genetic potential of each boar for economy of production and the carcass quality of his offspring.

Economy of production means the ability of the boar to transmit to his progeny the qualities of improved food conversion, rapid growth rate, and dressing percentage or 'carcass yield'. Carcass quality is based on the value of high lean-meat content in relation to the amount of bone and fat.

The performance of each boar is compared with other boars of the same breed that are being tested at the same time at the station. This is known as a contemporary comparison. Also, the performance of the castrate and gilt for economy of production is recorded, and their carcasses are dissected following slaughter at 90 kg liveweight.

Other characteristics, such as the boar's conformation, ability to walk well, soundness of feet, and so on, are also judged visually. The final 'mass' of information is analysed and classified with the aid of a computer.

The final results are given on a 'points score' basis—the higher the number, the better the boar is likely to be. The numbers range from ninety upwards, and any boar scoring less than ninety or being considered weak on his legs is slaughtered. The 'points score' system is used by farmers as a sound means of valuing a boar when making a purchase. In other words, the higher the score, the greater will be the sale price. Boars with a high performance rating offer a real opportunity to bring about improvement in your herd.

Breeding and rearing boars

If you have a particularly good sow of proven ability, and can mate her to a similar-quality boar, then you may consider rearing a boar pig as a future herd sire.

91

MLC MEAT AND LIVESTOCK COMMISSION

P.O. BOX 44, QUEENSWAY HOUSE, QUEENSWAY, BLETCHLEY, BUCKS.   Telephone: BLETCHLEY 4941

	FORM	TEST
	CT 21	8436/ 7

Report of Group Test 8436/ 7 **Born** 21ST SEPTEMBER 1970 **Tested at** CORSHAM **Report issued** 1ST APRIL 1971

TESTED ON BEHALF OF: **BRED :** LARGE WHITE

LIONEL ORGAN,
MANOR FARM,
SOUTHAM,
CHELTENHAM. GLOS.

Tel: CHELTENHAM:20236/20168
HERD.PREFIX:, SOUTHAM:

		TEST REPORT
Sire of Group	LEAFIELD KING DAVID 60TH	EAR No. HERD BOOK No.
Dam of Group	SOUTHAM FANNY 17TH	4196 93280 2426
Sire's Sire	LEAFIELD KING DAVID 2ND	368475
Sire's Dam	GRUNTYFEN FANNY 2ND	56904
Dam's Sire	MEPPERSHALL CHAMPION BOY 12TH	375750
Dam's Dam	SOUTHAM FANNY 2ND	22702

POINTS SCORE

EAR No. OF BOAR	6178	6181	AVERAGE OF BOTH BOARS	CONTEMPORARY AVERAGE
ECONOMY OF PRODUCTION	80	85	83	50
CARCASE QUALITY	87	96	91	50
TOTAL	167	181	174	100
BOAR CATEGORY	LICENSED	LICENSED		

NOTES :

OTHER INFORMATION
(DIFFERENCE FROM CONTEMPORARY AVERAGE)

CHARACTER	EAR No. OF BOAR		CASTRATE AND GILT
	6178	6181	
FOOD CONVERSION	0·178	0·318	0·15B
DAYS ON TEST (60-200 lb.)	3W	1B	2W
CONFORMATION SCORE		1B	1B
C (mm)	AVE	1B	3B
K (mm)	2W	AVE	2B
LOIN 2 FAT (mm)	1W	1B	6B
SHOULDER FAT (mm)	3B	1B	5B
KILLING OUT %			+1W
TRIMMING %			1·0B
LEAN % IN RUMP			AVE
EYE MUSCLE AREA (sq cm)			2·7B
STREAK SCORE			1W
AVERAGE LENGTH (mm)			81½

B : Better
AVE : Average
W : Worse

Fig. 22(b).   Boar test

It is somewhat difficult to select a young boar at say six weeks of age (castration time), but as a guide select a well-grown sucker, with fourteen well-placed supernumerary teats—this trait will later be transferred to his daughters.

If possible rear the pig with a litter brother, being either an entire or castrate. They are best housed in pens with an outside yard, where they may exercise and, if possible, see other pigs. Feed the boar as much food as he will clear up in about twenty minutes, twice per day. A grower-type ration should be used, see page 76.

On-the-farm testing

It is possible for breeders to test future breeding stock—either boar or gilts—on their farms simply by feeding them separately and recording the amount of food fed and the number of days from the time the pig is 20 kg until it reaches 90 kg liveweight. This will give the daily liveweight gain and food-conversion ratio.

The carcass quality may be indicated by the use of ultrasonic testing. The ultrasonic instrument—which is similar to radar equipment—measures the amount of backfat at the C and K measurements—this in turn indicates the amount of lean-meat content.

The M.L.C. operate an 'on-the-farm testing service'. They provide the use of ultrasonic testing equipment, and give the farmer advice on keeping the necessary records to produce a 'rating index'.

Licensing of boars

All boars which are intended for breeding must be licensed when they are six months old. A boar licence is granted after the boar has been visually inspected and passed by a Ministry of Agriculture livestock officer. The officer looks for a well-grown, healthy pig, typical of its breed, and pays particular attention to the boar's feet and legs. If the boar is satisfactory the officer will tattoo a Crown in the right ear. If the boar fails the inspection then the letter R is tattooed in the left ear and the boar must be castrated or slaughtered.

Should the owner object to his boar being rejected then he

PIG ACCREDITATION SCHEME

Interpretation of Test Results

INTRODUCTION

A group of pigs for testing consists of two boars, one castrate and one gilt all from the same litter. Four litter groups, all by the same sire, constitute a progeny test of that sire. An individual report is issued on each group and a final report is issued on the sire when four of his groups are tested.

The main object of the test is to assess the genetic potential of each of the young boars for economy of production and carcase quality.

The boars are assessed on what is known as a breed contemporary average: they are compared with other boars of the same breed tested at the same station at the same time. The average points score for economy of production and for carcase quality will always be 50, a total of 100 for the two. Consequently, individual boars can be compared with the average and with each other.

POINTS SCORE

In arriving at the points score (selection index) the fullest use has been made of expert guidance and past experience. It will be reviewed as further data from the scheme, and from the progeny of tested boars, becomes available. Each boar is assessed on his own performance, supplemented by the performances of the castrate, gilt *and other boar*, in his litter group. No further adjustment is necessary.

(i) Economy of Production

The points for economy of production are an assessment of the capacity of the performance tested boar to transmit to his progeny the qualities of food conversion, rate of gain and killing-out percentage.

(ii) Carcase Quality

The points for carcase quality are an assessment of the capacity of the performance tested boar to transmit to his progeny the qualities of lean as a percentage of carcase weight, distribution of lean and eye muscle area. (Assessment of a young boar for carcase characteristics depends to some extent on direct measurements, but to a greater extent on correlated factors, including the carcases of his litter mates.)

EXTRACT FOR PUBLICATION

The details within the bold lines are the minimum extract for publication. Any advertisement must include the points for the boar advertised or, if relatives are mentioned, the average points for both boars.

OTHER INFORMATION

Assessments of characters are given as so much better (B), the same (AVE) or so much worse (W) than the relevant contemporary average.

(i) Days on Test

This is based on the average number of days taken to grow from 60 to 200 lb. liveweight.

(ii) Conformation Score

This is a subjective assessment of the general shape of the carcases of the castrate and gilt. Scores are based on seven values, ranging from three points worse to three points better than average.

(iii) Fat Depths

C and K These fat measurements are taken at standardised points from the centre line of the back – by echo sounding the boars and from the cut face of the back rasher of the castrate and gilt.

Loin 2 and Shoulder The depth of the former is the minimum over the rump muscle and the latter is the maximum in the region of the shoulder – by echo sounding the boars and by measurement on the split carcase of the castrate and gilt.

(iv) Killing Out and Trimming Percentages

$$\text{Killing out percentage} = \frac{\text{Cold carcase weight} \times 100}{\text{Last liveweight}}$$

$$\text{Trimming percentage} = \frac{\text{Trimmed carcase weight} \times 100}{\text{Cold carcase weight}}$$

Cold carcase weight = weight of hot head and chine + twice the weight of the cold left side (including flare fat, kidney, feet and fillet).

Last liveweight = mean of the last three liveweights.

Trimmed carcase weight = twice the weight of the cold left side (excluding flare fat, kidney, feet and fillet).

(v) Lean Percent in Rump

This is the percentage of lean meat in the rear part of the back. This information is obtained from either the castrate or gilt.

(vi) Eye Muscle Area

This is measured on the castrate and gilt after cutting the carcases across at the head of the last rib.

(vii) Streak Score

This is a subjective assessment of the appearance of the streak. Scores are based on seven values ranging from three points worse to three points better than average.

LENGTH

The warm, hanging carcase is measured from the pelvic bone to the first rib, *after the head has been removed*.

PALE EYE MUSCLE

After a side from the castrate and gilt has been cut at the head of the last rib it is judged subjectively for pale and wet muscle. If this condition is observed rarely it may be ignored. But if it occurs frequently, seek advice.

LICENSING

Boars are normally licensed on the testing stations. Boars scoring 89 points or below are slaughtered. Any boar will be slaughtered if it is unsatisfactory on its legs.

CASUALTIES

If there are casualties in a litter group, the report is so adjusted that the results can be interpreted in the normal way.

EVALUATION

The 'Points Score' is given as the means of evaluating potential breeding stock. The 'Other Information', which has already been taken into account in the 'Points Score', is provided as a fuller explanation.

Fig. 22(c).   Accreditation scheme

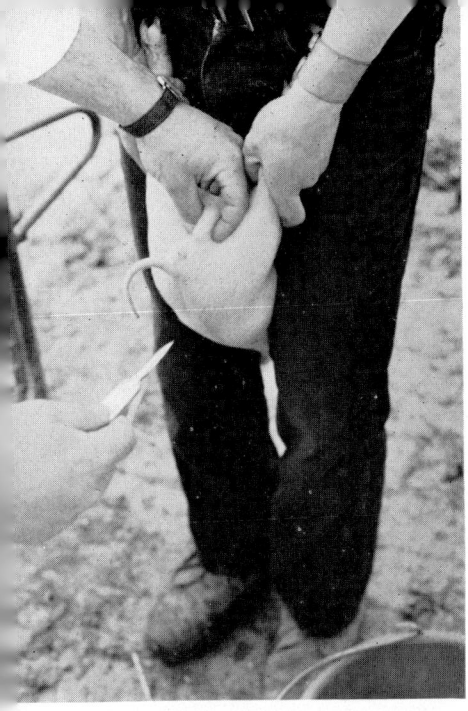

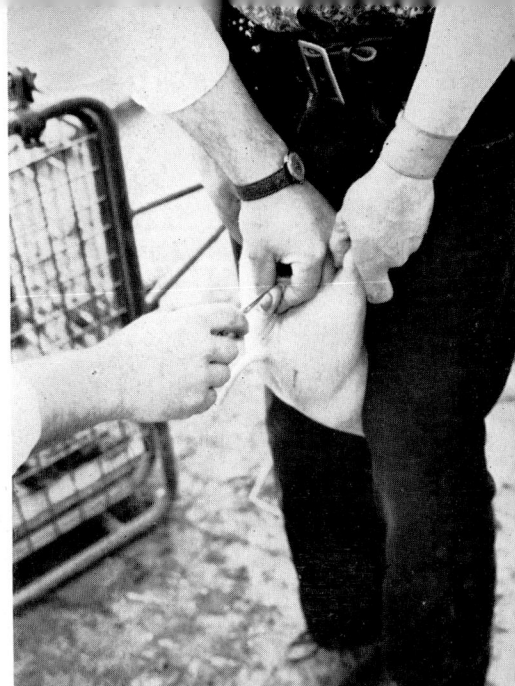

After cleaning the scrotum with surgical spirit grasp the testicle firmly between first finger and thumb

Make a bold incision with the scalpel. The testicle will 'pop' forward

Photographs by C. S. Freeman

After removal of the second testicle dust the cavities with sulphanilamide powder to prevent infection in the wound

Cut the spermatic cord to remove the testicle

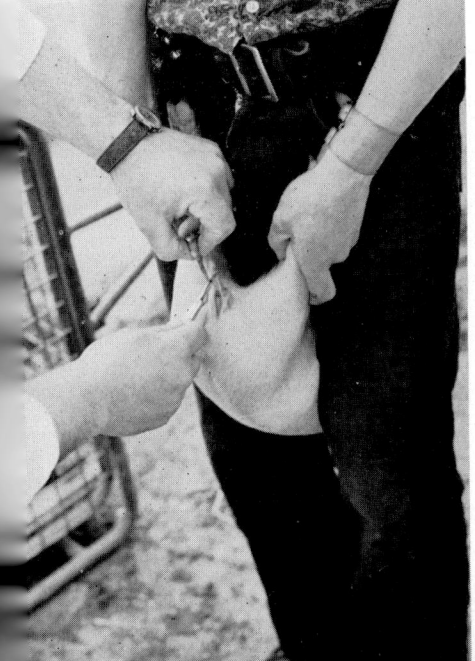

Cage rearing

has the right to appeal, and an independent pig expert will assess the boar's merit.

Application for a boar licence must be made when the pig is five months old—the fee is £1·75 and application forms may be obtained from the Ministry Divisional Offices.

The existing boar-licensing scheme is open to some criticism and is at present (1972) under review by the Ministry of Agriculture.

Boar management

Although young boars become sexually mature when four to five months old, it is unwise to allow them to serve gilts until they are at least seven to eight months old and weigh around 100 kg.

At this stage they may be allowed the occasional service—say once per week—but their work should be restricted until they are around twelve to fifteen months.

They may then serve a sow per day for short periods, with rests in between.

A mature boar can cope with up to forty sows per year, and thus make in the region of ninety to a hundred services—the odd sow returning to heat and some sows may be served twice.

However, if you have forty or more sows you will probably wish to keep two boars in case one should go lame or be ill. Also, we must remember that if batch farrowing is adopted then there may well be several sows on heat on the same day.

Boars should always be housed in strong pens with a generous amount of exercise room, and preferably where they can see some sows. They should be fed well in order to keep them in fit condition, but care should be exercised to prevent them from becoming overfat, which may impair their fertility.

Twelve Management of Gilts from Six Months Onwards

If you decide to select your breeding gilts from the bacon pens when they are around 90 kg l.w. it will be necessary to give them some form of exercise in order to strengthen their legs and harden the soles of their feet before they are served at around eight months. This can be achieved by running the gilts in the open yard, with some shelter for sleeping quarters. The gilts will soon become acclimatised to the open conditions, and will greatly benefit from the fresh air and exercise.

Where possible, they should be fed from individual feeders, to prevent bullying. Each gilt will require 3 kg of sow and weaner meal per day to keep her growing without becoming over-fat.

During the summer months gilts may be run out of doors at pasture or in woods. If good grazing is available, the meal allowance may be reduced to 2 kg per gilt per day.

Outdoor systems reduce the stress induced by overcrowding and lying on concrete, the risk of airborne diseases, etc., and offer the advantages of direct sunlight, vitamins and minerals from the soil, fresh air and exercise. There is no doubt that gilts managed in this way are fitter and healthier than gilts that are always kept indoors.

Signs of heat

Gilts become sexually mature at between five and six months, and will, if allowed, accept the boar at this age. However, it is most unwise to breed from gilts before they are $7\frac{1}{2}$–8 months old and weight 110–120 kg. Early matings cause a great strain on a gilt and may seriously affect her growth and health.

Gilts come on heat every 20–21 days. The heat period

normally lasts from 12–36 hours, during which time the gilt will accept the boar. The signs of heat are restlessness and a marked swelling of the vulva. With white breeds, the vulva changes to a pinky red colour, which makes detection quite easy.

A good indication of when the gilt is ready for mating is to place your hands across the loin and apply pressure. If the gilt stands still, then she is ready for the boar.

Mating

Ideally the gilt should be served twice, and then housed away from the other pigs for a day or two in order to gain high conception rates. The gilt should also receive twice her normal ration on the day of mating and the day after. During the heat period the ovaries produce 18–20 eggs which are fertilised. After conception takes place the fertilised eggs attach themselves to the wall of the uterus to develop into foetus.

Unfortunately, many of the fertilised eggs will be re-absorbed into the sow's bloodstream if she is upset, knocked about by moving from place to place, or bullied by other sows. It is for this reason that the gilt is best housed separately for a couple of days after mating and fed well.

Service

Unlike cattle and sheep, the act of copulation in pigs is rather lengthy. A boar may take from five to twenty minutes to serve a sow.

If a large heavy boar is used it may be necessary to put the gilt in a service crate, where the boar rests his front feet on a raised floor.

Alternatively, small gilts may be stood on sloping ground, facing uphill, for large boars, and conversely large sows may be stood facing downhill for mating with young boars.

Conception rates in pigs are usually fairly high, but a careful watch should be made 20–21 days after service for gilts 'turning to the boar'. Gilts and sows that return more than twice are best sold for slaughter.

Pregnancy

Pregnancy lasts a little under four months, and during this time the gilt should receive adequate food to keep her growing and provide nutrients for the developing litter; 3 kg of sow and weaner meal should be adequate, although a little extra may be required in winter if the gilts are running out of doors.

As pregnancy advances, the gilt will become more docile, and the abdomen will greatly enlarge. The udder will develop markedly during the last month.

Summary

Gilts should not be served until they are $7\frac{1}{2}$–8 months and weigh 120 kg l.w. for bacon breeds, or 110 kg for dual-purpose breeds.

Heat cycle every twenty-one days—swollen vulva
restless

Have gilts served twice if possible.

Feed extra food on the day of service and the following day.

Check gilts three weeks after service for those who 'return to service'.

Allow 3 kg sow and weaner meal per day during pregnancy.

Keep gilts fit by allowing them plenty of fresh air and exercise.

Thirteen   The Sow and Litter

Preparation for farrowing

Warm, dry, comfortable quarters are the key to success at farrowing time. The pen should be thoroughly cleansed and disinfected well in advance of the farrowing date. Electric fittings should be checked and infra-ray lamps made available.

Sows should be introduced to their pens about 2–3 days before pigging, and gilts given a little longer period, say 4–5 days. This is especially important when farrowing crates are used, because a gilt may become upset at finding her movements restricted.

In this case it is wise to put the gilt into her crate for 2–3 hours, then let her out for the first day or so.

If the weather is warm, the sow may be washed all over with warm soapy water to remove dirt, lice and roundworm eggs before she is brought in. In cold weather brush the sow with a stiff dandy brush to remove the dirt, then dust the pig with an insecticide powder. The udder is best washed and dried as quickly as possible.

The sow's diet should be changed gradually from sow-and-weaner meal to 2–2·5 kg of coarse bran, fed in a wet mash. This will keep the bowels open, thus preventing impaction of the rectum with hard faeces which will cause the sow undue stress at pigging if not removed.

Should a sow become constipated, a mild enema should be administered.

Enema equipment: one clean bucket, ten litres of clean warm water, one enema pump—vaseline—50 g soft soap.

Dissolve 50 g soft soap in bucket of clean warm water (38°C).

Apply a little vaseline to the rubber tubing to act as a lubricant when inserted into the sow.

Gently force tube into rectum about 300–350 mm.

Slowly inject the soapy water into the rectum then remove tube.

Make sure that the expelled water and dung drains away from the sow's bed.

Fig. 23(a).   Washing sow

Signs of farrowing

During the last days of pregnancy the sow will become increasingly restless. The vulva becomes greatly enlarged and the udder appears to be full. The sow will chew up straw and make herself a 'nest'. About 12–24 hours before she pigs the teats become turgid and you can draw milk at the slightest touch.

At this stage switch on the infra-ray lamps in the creep to raise the temperature in the pen to as near 27°C as possible. Bright emitter lamps are preferable, as the light will attract the piglets away from the sow, and reduce the risk of overlaying.

100

Bedding

Short chopped straw, sawdust or wood shavings may be used for bedding. Avoid long straw, since new-born piglets will have great difficulty in walking through it.

Farrowing

The sow will lie down when farrowing, and normally she grunts and lifts her hind legs as the labour pains become more severe. Generally speaking, sows have little difficulty with pigging, and provided adequate preparations were made (warm dry bed— infra-ray lamp switched on in the creep) she will farrow successfully without interference from the pigman. Sows vary considerably in the time they take to farrow: some sows will pig in a couple of hours, whilst others will take eighteen hours or more. The majority of sows take around 4–6 hours.

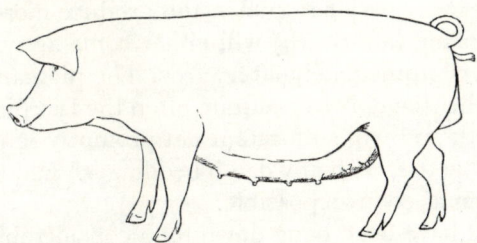

Fig. 23(b).   Gilt near farrowing

The sow will strain for some time before the first pig is born, but after this she seems to find parturition much easier. Quite often a sow will rest for an hour between the first and second pig, and then deliver the rest of the litter quickly.

As each pig is born, it severs the navel cord, and then quite miraculously walks round the sow's hind leg to find the udder, so that within minutes of being born the young pig is sucking milk.

Occasionally a sow becomes upset at farrowing, and will continually get up and down, often stepping on her pigs and making it difficult, perhaps impossible, for them to suck. In this case the pigman may catch each pig by the hind leg and place them in a deep box lined with hay. The box may be placed under an infra-ray lamp to dry the piglets and keep them warm.

Fig. 23(c).   Sow in farrowing crate

The piglets should be returned to the sow when she has finished pigging and settles down. The strongest piglets are best placed on the rear teats and the weaker ones on the forward teats, as these are easier to suckle and produce more milk.

The cleansing (afterbirth) will either come away as the last pig is born or immediately afterwards. The pigman should be present at this stage, because quite often the last pig is born in the afterbirth and will suffocate if not promptly removed. The cleansing, together with any damp bedding, should be removed from the pen as soon as possible.

Sows that have been lying down for a considerable time, say 18–20 hours, should be encouraged to stand up and move about. This will usually make them urinate and empty their bowels. If a sow is down too long there will be a risk of the stale urine in the bladder causing ill health.

A warm feed of bran should be given to the sow after parturition, and then she should be left undisturbed for the next 12–24 hours.

Critical three days

The newly born piglet is susceptible to many dangers during its first few days. With little or no hair for protection, it is particularly susceptible to colds and chills. It may be trodden on or crushed by a clumsy sow, or starved of food should a sow's milk supply fail.

Watch the litter's progress carefully, and if the piglets are squealing and hungry, carefully examine the udder. Should it feel hard, and either very hot or extremely cold, send for a veterinary surgeon at once.

By the third day, the 'danger period' will be over, and the pigs should now make rapid progress.

De-tusking

Baby piglets have needle-sharp teeth which can cause the sow discomfort when the pigs are sucking. The tips of the canine and pre-molar teeth should be removed, using a sharp pair of teeth pincers.

Piglet anaemia

Iron constitutes about four parts in a million of an animal's bodyweight, and is an essential constituent of haemoglobin, the pigment in red blood corpuscles which is responsible for carrying oxygen throughout the body.

A deficiency of iron causes anaemia, and this is quite a common disease in young pigs reared indoors. The sow's milk contains very little iron, and so additional iron must be administered to the newly born piglet either orally or by injection.

Anaemia can attack baby pigs in two ways: acute, which usually causes sudden deaths in the strongest pigs, and chronic anaemia, which causes a general unthriftiness and scouring.

The acute form is often called 'Thumps', owing to the piglet's rapid breathing as the heart attempts to supply the body with oxygen. Chronic anaemia is characterised by a general unthriftiness, paleness of skin and profuse scouring. If left untreated the pig will either die or become a 'poor doer'.

Prevention and treatment

Iron capsules, or pastes, may be administered orally by placing on the pig's tongue. Alternatively, an injection of ferrous sulphate may be made.

Placing a sod of earth or small amount of coal slack in the creep for the piglets to 'rout' in will often prove satisfactory.

103

Feeding the suckling sow

For the first few days after farrowing the sows should be fed lightly on a mixture of equal parts bran and sow and weaner meal. After about three days the ration should be changed to sow and weaner and increased according to the number of piglets suckling. As a rough guide, the sow should be fed 3 kg sow and weaner meal for maintenance, plus 0·25 kg of meal for each suckling pig. For example, a sow with ten piglets will receive:

$$3 \text{ kg sow} = 3 \quad \text{kg}$$
$$\text{Plus } 10 \times 0\cdot25 = 2\cdot5 \text{ kg}$$
$$\overline{}$$
$$5\cdot5 \text{ kg}$$
$$\overline{}$$

Obviously you must watch the sow's body condition, and feed more meal if she is losing condition too quickly. Sows with fourteen or more pigs are best fed *ad lib*. It is most important that only fresh, appetising, properly balanced rations, with adequate vitamins and minerals, are fed. Clean water is, of course, essential and must be freely available. Suckling sows will greatly benefit from ten litres of skim milk per day if available. Turning sows out to pasture for an hour or so per day will, if practicable, also benefit the lactating sow.

Creep feeding is a standard practice and should be started when the piglets are 5–7 days old, although it is unlikely that they will eat much meal before they are two weeks old. A soundly constructed creep is essential to prevent the sow from getting at the piglets' ration. Fresh pellets or meal and clean water should be provided daily. Troughs or self-feed hoppers may be used.

Weighing at three weeks

This is strongly recommended, since this is a real opportunity of measuring the sow's milking ability. Remember that the weight of the litter at three weeks depends almost entirely upon the performance of the sow in producing the litter and feeding it. The M.L.C. figures show an average weight for litters at three weeks to be around 55 kg This should be considered the

104

minimum acceptable weight for sows whose progeny are to be kept for future breeding stock.

Fig. 24(a).   Weighing three-week-old pig

Ear marking

It is convenient to ear-tattoo pigs at three weeks when they are being weighed. If ear notching is practised then this is best done when the piglets are a few days old. Ear tags, however, should not be used until pigs are weaned, since the pigs may bite at the tags when suckling, and this may lead to other vices (see page 118).

Castration

Many pigmen castrate the male pigs before they are two weeks old, although traditionally they are cut at four weeks. Castration prevents incestuous breeding and is necessary to prevent taints (unpleasant flavours) in the pigmeat. The earlier the operation is performed, the better, since, if delayed, the pig may suffer a severe setback (see page 117).

Weaning

In order to achieve two litters of pigs per sow per year, farmers have weaned pigs at eight weeks old. With improved housing and management today there is no need to leave pigs on the sow for that period, and they may, quite safely, be weaned at six weeks, when they weigh around 13 kg liveweight. The important points to remember are that the sow must be removed from the pigs and not *vice versa*. In this way the pigs remain in familiar surroundings, which, of course, reduces stress. It is also important that the piglets should continue on their creep feed ration for at least two weeks and when their diet is changed it must be done gradually.

If this procedure is adopted, the weaning check will be minimal and the piglets should reach 18–20 kg by eight weeks. They can then be changed to sow and weaner ration and if necessary moved into a weaner pool.

Early weaning

Early weaning at three weeks of age has been advocated by feedingstuff manufacturers for many years now, and there is no doubt that the system increases the output per sow.

On many farms early weaning has proved to be highly successful and indeed profitable to the farmers, yet on other farms the system has been abandoned owing to outbreaks of scouring which leads to poor growth.

Briefly, the system requires clean, hygienic, warm buildings, fresh, specially formulated, feedingstuffs and expert management. One way of achieving the housing requirements is to build straw bale pens out in the open, on clean pasture. The piglets are removed from the sow on the 21st day after birth and placed in the straw hut. They are fed proprietary 'sow milk' replacement meal and clean water *ad lib*. Later the piglets are gradually changed on to a creep ration. Reared in this way the piglets usually reach 16–18 kg by eight weeks, when they can be put into a weaner pool.

The sow will require special management. She should be fed only 3 kg meal per day during the three-week suckling period, otherwise she may become overfat.

She should be served on the first heat period after weaning, and in this way will produce five litters in two years, instead of the usual four.

Caution

It must be emphasised that 'early weaning' pigs require

exacting standards of husbandry. The beginner is advised to adopt six- or eight-week weaning until he gains experience.

Worming

Roundworm infestation is a common complaint in pigs. Young pigs are particularly susceptible, and so all weaners should be wormed before they are moved to the weaner pool.

The life-cycle of the roundworm, symptoms and control, is dealt with on page 183.

The dry sow

The dry sow should be housed separately for a few days after weaning, and until her udder dries off. Normally, sows will take the boar 4–7 days after weaning.

The served sow may be returned to the dry sow quarters 2–3 days after service.

Summary

Prepare warm, dry, comfortable quarters for farrowing—check electric fittings—use short straw or wood shavings and sawdust for bedding—feed sow laxative diet—introduce gilts to pen 4–5 days before farrowing.

Sows best left alone when farrowing—after pigging remove the cleansing—feed sow, then leave undisturbed.

Third day: Detusk and inject with iron, or give oral dose. Gradually increase sow's feed.

7–10 days: Offer creep feed to baby pigs.

Three weeks: Weigh litter, ear-mark, and castrate males.

Six or eight weeks: Wean litter by removing the sow and allowing the pigs to stay in their pen for a few more days.

Worming: Dose the litter for roundworms, and then mix them in the weaner pool.

Dry sow: House separately after weaning until she dries off—will usually take the boar 4–7 days after weaning.

Fourteen Cage Rearing

The revolutionary idea of rearing baby piglets in wire cages was first demonstrated in Germany by *Heinrich Biel* in 1954. The Biel system involves weaning the piglets at 2–4 days old and keeping them in individual cages where they are fed on a specially formulated liquid feed.

Cage rearing is now being developed on the Continent and there is much interest in dry feeding—a new system which has been demonstrated by *Dr. H. Van der Heyde*, at Ghent University, Belgium.

In this country the system is being investigated by research scientists and their evidence to date would suggest that cage rearing could become commercially sound within the next few years. There are at present several large commercial units in Belgium, Germany, Holland, Italy and Switzerland.

The Belgian system is to remove the piglets from the sow when about 4–10 days old. The piglets will have received colostrum and should weigh about 2 kg. The wire cages are usually built in three tiers, so that it is possible, where sows farrow in batches of three, to remove a third of the heaviest piglets from each sow at four days, and place them in the bottom cage, then a further third at six days, and complete the final weaning at ten days. This will give all the piglets a good start on sow's milk and should result in rearing three cages of evenly sized pigs.

Once removed from the sow, the piglets are offered specially formulated milk-replacement pellets and water. The piglets are fed twice daily and disturbed as little as possible between meals. On the Continent it is recommended that the piglets should be kept in total darkness between meals, as this will prevent vices such as navel sucking and tail biting.

Obviously, great attention must be paid to the environment, the temperature, humidity, and air movement should be kept constant, and cages must be kept scrupulously clean.

Dr. Van der Heyde recommends an air temperature of 27°C for newly caged piglets which is gradually lowered to 22°C when the piglets weigh around 18 kg. The relative humidity should be between 40–60%. Below this coughing may occur and above 60% problems of scouring have been found. Tranquillity is also important, because every time piglets are disturbed in their normal sleeping period they tend to react by biting each other. To this end, keeping the piglets in darkness between their feeding times is recommended.

With regard to hygiene, most rearers appear to have the cages steam cleaned between batches and then rest the cages and rearing house for several weeks.

Benefits of cage rearing

Should cage rearing become a commercially sound business in the future, one can envisage many advantages and other changes in traditional pig management, for example:

1 Gilts may be bred from at a much earlier age if they are not expected to rear their litter for a long period. In fact it may be possible to mate gilts when around five to six months old, farrow them at nine to ten months and then sell them as a meat animal. Pregnant gilts are extremely efficient converters of food into developing pigs and putting meat on to their bodies. If such a system were practised one could use a gilt as a breeding animal and heavy hog.

2 With normal breeding stock there is the advantages of feeding less meal to the sow per annum by the saving of two lactations and the fact that theoretically it should be possible for a sow to produce three litters of pigs per year.

3 Housing costs are considerably reduced, as the sows only occupy the farrowing crates for about two weeks and the cost of cages is considerably less than 'follow on' accommodation for suckling sows. However, it must be remembered that the running costs, especially heating, will be more expensive with cage rearing.

Conclusion

Although it is claimed by enthusiasts that cage rearing is 30% more efficient than conventional rearing and the fact that some breeders are averaging 2·6 litters per sow per annum, it must be remembered that cage rearing demands the highest possible standards of hygiene, stockmanship and management. In fact one could say that the pigman requires a 'clinical approach' and the pigs must be kept in 'hospital-like conditions'.

Fifteen   Pig Handling

No one could describe the pig as being an intelligent or affectionate animal. They are in fact rather nervous and inclined to be mischievous and destructive. Nevertheless, if approached quietly and firmly pigs will soon respond to handling and become docile. This is particularly important with breeding stock. A good pigman should be able to walk amongst his stock with the minimum of fuss and disturbance.

Handling and catching

Small pigs may be easily caught by grasping either of the hind legs just above the hock joint and lifting them off the floor. Stronger pigs should be caught by grasping them behind the shoulders (see Fig. 24(b)), using your outstretched hands. In this way pigs up to about 50 kg may be handled with little difficulty. Heavier pigs should be run into a cage—such as the pig scales— or restricted by using a rope.

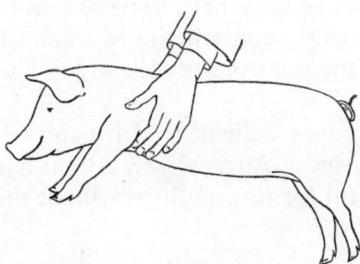

Fig. 24(b).   Catching a young pig

Always try to 'corner' your pigs by driving them into a small pen or placing a piece of corrugated iron or boarding across the

111

corner of the pen. Pigs are extremely agile and can turn very quickly, which makes catching difficult if they are not in a confined space.

Driving pigs

Adult pigs are generally fairly easy to move from place to place, provided the pigman approaches the task quietly and firmly. A 'board' measuring 0·70 m × 0·70 m × 10 mm with a suitable handle and a 'bat' measuring 1 m × 75 mm × 10 mm is most useful for guiding the pig. The 'board' should be used to simulate a wall, thus preventing the pig from looking from side to side.

Fig. 25.   Driving pigs

The 'bat' should be used only to 'tap' the pig lightly should it refuse to move forward.

Where a number of sows have to be moved fairly frequently, then it is worth while constructing a solid wall 'race' or pig walk. As long as the pig can see a clear way ahead it will move forwards.

Small pigs are more difficult to drive and wherever possible a 'race' should be used. Alternatively a light hand trailer or pig scales can be useful for moving litters about the farm.

Injections

There are two ways in which you may inject a pig: (a) subcutaneously, which means under a loose fold of skin and (b) intramuscular, which is to inject directly into the muscle.

112

Subcutaneous

The most suitable site for injecting under the skin is at the base of the ear, where loose folds of skin are clearly available. This site also has the advantage of being in a 'clean area' of the pig's body. An alternative site is in the groin region.

Clean the site with surgical spirit and then inject under loose skin, which may be held between the first thumb and finger of your left hand (see Fig. 26(a)).

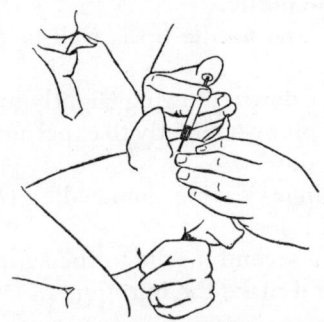

Fig. 26(a). Subcutanous **injection**

Intramuscular

This injection is usually made at the base of the neck with adult pigs and in the fleshy part of the hind leg with young baby pigs, for example iron injections.

The site is first cleansed with surgical spirit and then the injection made direct into the muscle (see Fig. 26(b)).

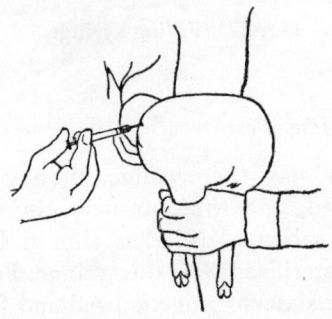

Fig. 26(b). Intramuscular injection

113

Use of hypodermic syringe

1 Assemble the syringe and needle; shake the bottle and swab the cap with clean surgical spirit.

2 Draw into the syringe a volume of air slightly more than the volume of liquid to be withdrawn, and thrust the needle through the rubber cap of the bottle.

3 Turn the bottle upside down, push the plunger to inject the air in the syringe into the bottle. If you do not do this, you will have difficulty in withdrawing the dose, as a partial vacuum will form inside the bottle.

4 Make sure that the needle tip is well below the surface of the fluid.

5 Pull the plunger down, drawing slightly more liquid than is required. Push the plunger slightly to expel any air bubbles, and adjust to the right dose.

6 Detach the syringe, leaving the needle in the cap for withdrawing subsequent doses.

7 Finally, attach a second needle to the syringe, expel any air from the needle, and make the injection.

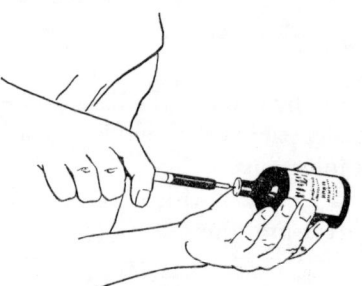

Fig. 27   Filling syringe

Care of syringe and needles

Immediately after use the syringe should be dismantled, thoroughly cleansed, and then sterilised by boiling in clean water for twenty minutes. Needles should be changed frequently, and then sterilised with the syringe. For intramuscular injections a 30 mm sixteen-gauge is used and for subcutaneous injection a 15 mm eighteen-gauge.

114

Note

1 Always discard part-used bottles at the end of the day.
2 Always check the dose on the bottle label and make sure the correct amount will be delivered by the syringe.
3 Do not inject pigs within four weeks of slaughter.

Nose-rings

Ringing pigs is an age-old practice that is still adopted on many farms in the belief that not only does it prevent pigs from routing up pastures, but that fattening pigs are quieter and less likely to damage troughs or attempt to unhinge doors with their snouts.

In olden times the village blacksmiths would force a piece of wire through the nostrils and then twist the ends together with a pair of pliers. Today we can purchase two types of pig ring— the self-piercing copper wire ring, which is most suited for growing stock, and the small 'bull ring' which is recommended for boars and sows.

To ring store pigs an assistant is required to catch and hold the pig in a sitting position and the mouth is held firmly closed see Fig. 30.

Using the special pig pliers, insert a copper wire ring and then insert the ring in the thick part of the snout. It is best to use at least two and preferably three rings per pig in case one should come out.

Bull-rings

To ring adult pigs you must first restrain them with a nylon

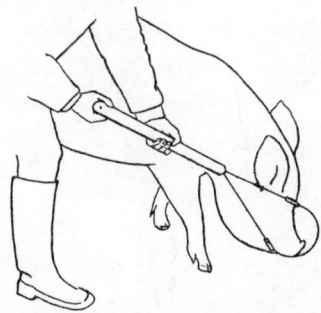

Fig. 28. Pig restrainer

115

rope or special pig restrainer. The rope is 'looped' and then held in front of the pig's nose until it opens its mouth.

The 'noose' is tightened firmly and then the end tied to a post. The pig will squeal and pull backwards and in so doing further tighten the cord.

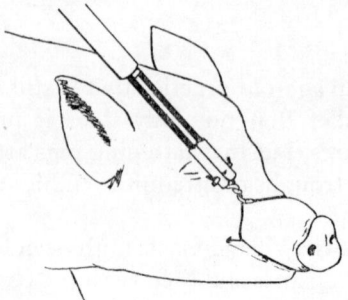

Fig. 29.   Pig restrainer in place

The self-piercing bull-ring, which is approximately 40 mm in diameter, is fixed between the inside of the two nostrils. A small screw is then used to hold the sides of the ring together.

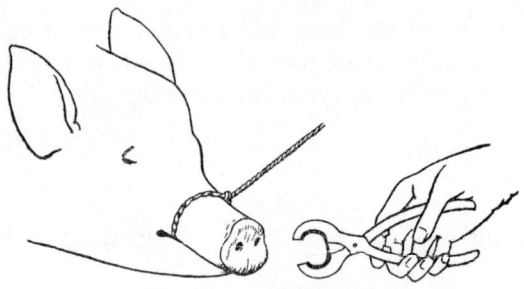

Fig. 30.   Ringing pig

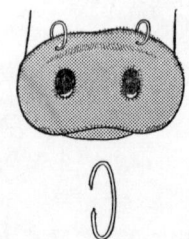

Fig. 31.   Copper ring

116

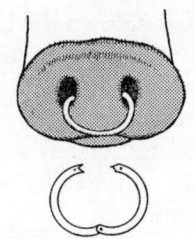

Fig. 32. 'Bull ring'

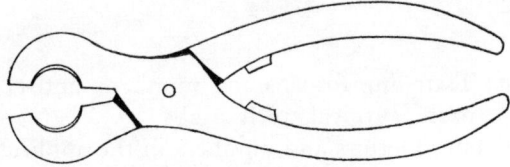

Fig. 33. Ringing pliers

Castration

Procedure

Again, it must be stressed that the beginner should receive personal instruction from a veterinary surgeon or skilled pigman before he attempts to use a scalpel.

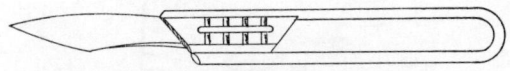

Fig. 34. Surgical scalpel

Equipment: One sterile scalpel and sterile blades, surgical spirit, cotton-wool cloth and bucket of warm water with mild antiseptic solution added, sulphanilamide powder.

Remove the sow from the litter.

An assistant should catch the male pig by the hind legs and hold it between his legs in an upside-down position.

Wash the scrotum with the warm water and antiseptic to remove dirt and dung.

Swab the scrotum with surgical spirit.

Hold the testicle between first finger and thumb by pressing the testicle forward with the third finger.

117

Make a firm, bold incision down the length of the testicle. At this stage the stone will pop out and may be gripped with the finger and thumb and pulled out, or the cords may be severed with the scalpel.

Repeat with the second testicle.

Dust the cavities with sulphanilamide powder to prevent infection of the wound.

Return the pigs to a clean, disinfected pen, with plenty of fresh bedding.

Ear tattoo

Equipment: Tattooing forceps and numbers, antiseptic tattoo paste, surgical spirit swabs.

Place the tattoo letters and numbers in the applicators.

Check that the letters are in the correct position by marking a piece of cardboard before tattooing the pig.

Pick up the piglet and swab the outside of the ear with surgical spirit swabs.

Tattoo the ear by piercing the back of the right ear with the tattoo set.

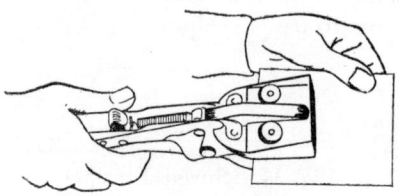

Fig. 35(a).   Ear tattoo

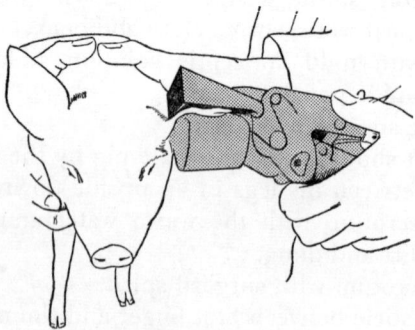

Fig. 35(b).   Ear tattoo

118

Rub black antiseptic tattoo paste into the markings with your thumb or an old toothbrush.

Tattoo the inside of the left ear.

Ear notching

This is best done when the piglets are three days old, and may be carried out when the pigs receive their iron treatment. Notching consists of clipping out a piece of the ear, which indicates a number, for example

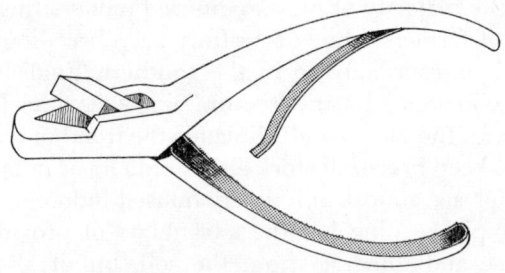

Fig. 36.   Ear-notching pliers

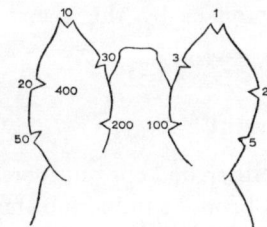

Fig. 37(a).   Ear-notching system

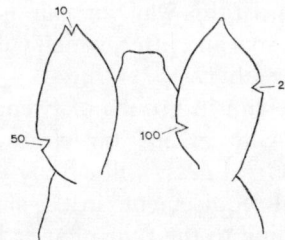

Fig. 37(b).   Example of notching Number 162

119

Sixteen Outdoor Pig Keeping

Although the majority of pigs are housed indoors there are still a number of farmers who successfully keep breeding pigs outdoors. This is especially so in the southern English counties where the climate and some free-draining soils offer favourable conditions. In the North and Midlands the trend in recent years has been to keep breeding stock either indoors or in open yards. Feeding pigs are almost universally housed indoors.

Outdoor pig keeping has the advantages of providing fresh air, exercise and minerals from the soil, but it also has the hazards of rain, snow and mud in the winter. Although there may be considerable savings in the cost of food and housing, today this is largely offset by the rising costs of land and labour.

Breeding management

Breeding sows may either be kept outdoors during their gestation period and then brought indoors to farrow or spend their entire lifetime outdoors. With the former system they may farrow in any month of the year, but if the latter system is adopted it is best to farrow the entire herd during the spring and autumn months. In this way you can use cheap range huts for protection of the sow and litter, which will be quite adequate during the milder weather.

The sows may be run in groups of twenty per hectare, and are best if rotationally grazed on clean pastures to avoid parasitic infection. Good grass will supply up to two-thirds of the sow's nutritional requirements in the summer months, but this will vary according to the type of sward and stocking rate. Sows should be 'rung' with a small 'bull-ring' to prevent them

120

from damaging the pasture by 'rooting' and this also avoids damage to huts, gates and fences.

In some areas, however, where sows are used to help with reclamation of old pasture or bracken-infested land, the sows are left unrung and allowed to tear up the ground as they search for the nutritious plant roots.

Fencing

Never forget that the pig has powerful shoulders and a natural capacity to escape. Strong fences are vital if proper control is to be maintained. Woven-fence wire netting is most effective, provided it is kept tight and strong straining posts are used.

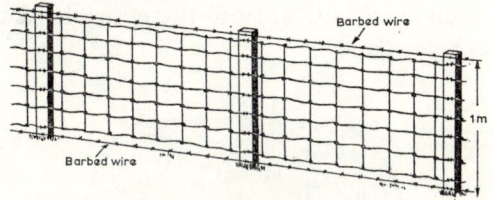

Fig. 38.   Pig-netting fence

Electric fencing may also be used for in-pig sows, but it is not so effective with sows and litters. Two wires are necessary, and these must be kept taut and clear of blades of grass or weeds that are likely to 'earth' the fence.

The fence unit may be battery or 'mains' operated. The battery operated unit is portable and so can be used in various fields, but where the fields to be fenced are near to the farm buildings a single mains unit can electrify all the fences and thereby save the cost of buying extra units.

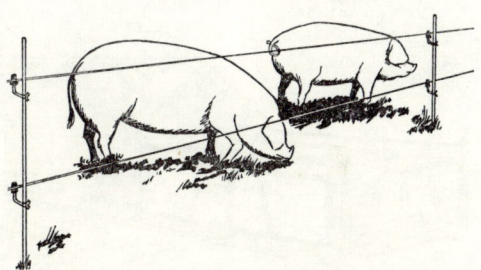

Fig. 39.   Two-strand electric fence

Accommodation

Outdoor housing falls into two types: the inexpensive non-insulated range shelter (Fig. 40) and the more expensive fully insulated hut (Fig. 41).

Range shelter

Provided the sows are farrowed during the spring and autumn months—say March–April and September–October—corrugated-iron range shelters are perfectly adequate for protection of the sow and litter. The shelters are extremely strong and durable and will last for many years with reasonable care and maintenance. The only disadvantage is that each sow will require an individual hut, because the entire herd will farrow in a 6–8-week period.

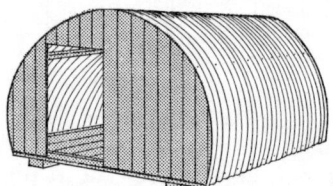

Fig. 40.   Pig shelter

Insulated hut

If you decide to farrow outdoors during the colder winter months and the warm midsummer period, then fully insulated huts will be required. These are expensive to build and because wood is the main material are not likely to last as long as the iron shelter. They can, of course, be moved from the fields and

Fig. 41.   Pig hut

122

stood on concrete in yards in the winter, but this is rarely satisfactory. Problems of carrying water in frosty weather to suckling sows make the system impracticable.

However, there is no doubt that in the colder northern districts the insulated house is superior to the shelter and allows farrowing over a longer period. There is the added advantage that one insulated hut can accommodate the needs of two sows—i.e. four farrowings per hut per year.

Type of sow

There is little doubt that the coloured breeds or their first cross are far superior for outdoor pig keeping than the white-bacon breeds.

The British Saddleback is numerically the most important 'outdoor pig'. It is essential that you obtain docile strains of sows that will forage well, are quiet at farrowing, respect fences, and produce vigorous pigs.

The routine day-to-day management is basically the same as described in Chapter 13. The boar is run with up to twenty sows at mating time, and may be kept with the sows until they are fairly heavy in pig. With good grazing in the summer only 1 kg of the large 'Jumbo'-type nuts will be required per sow per day. These are best fed once per day and scattered fairly widely over the ground. In the winter months 2 kg or possibly $2\frac{1}{2}$ kg of nuts may be necessary to maintain the sow in good body condition.

As farrowing time approaches, check that sows are sleeping in a hut and provide them with suitable short bedding material.

At farrowing you can shut the sow in her pen, and then it is best to leave her alone. Remember that sows kept outdoors are likely to be more nervous than those kept indoors with extra handling. Therefore they are more likely to become upset if you are frequently 'looking in' than if you leave them completely alone.

Once farrowing is over, the sow will soon take her litter outside. If you can confine her in a small hurdled pen for a week or so this will be advantageous, the little pigs are less likely to get 'lost' and you can ensure that the sow is milking well. After a week the sows' litters can run with one another, although this may lead to cross-suckling if a particular sow is short of milk. Alternatively, the sow may be tethered.

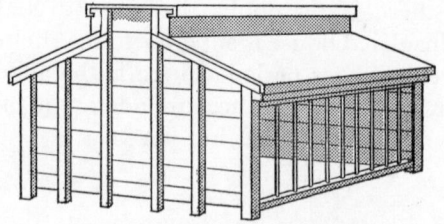

Fig. 42.   Communal creep

A strongly constructed communal creep with a self-feed hopper capable of holding an adequate supply of creep-feed meal or 'crumbs' should be available for the piglets when they are around two weeks old. You will also need some form of holding pen to catch the males for castration when about three weeks of age. This may easily be constructed with several pig hurdles.

Fig. 43.   Pig hurdles

Fig. 44.   Portable pen

The pigs should be weaned when around 18–20 kg. This will normally be when the pigs are about eight weeks old. Some farmers allow the pigs to suckle until ten weeks, but this should be avoided, especially if the sow's body condition is low, as it may result in difficulty in getting the sow back in pig. The weaners are then housed in a suitable 'weaner pool'.

Conclusion

Where inexpensive free-draining land is available, and the climatic conditions are favourable, outdoor pig breeding can be worth while. It is at times demanding in labour, but there is no doubt the system produces very fine weaner pigs.

Seventeen Weaning to Slaughter

Bacon pigs

One of the secrets of successful livestock production is to keep animals growing without check from birth to slaughter. Nowhere is this advice more important than in pig production. Healthy pigs that grow quickly have better food conversion and carcass quality than pigs that receive checks. The aim should be to achieve maximum growth rate in the early stages and then slightly less in the final stages, and to achieve good food-conversion ratios consistent with high carcass quality.

Weaner pool

In larger herds, or where 'batch farrowing' is practised, it is usual to mix weaners together in lots of twenty to fifty when they are eight to ten weeks old. The pigs are fed from self-feed hoppers and water bowls until they reach around 45 kg. They may be housed in the bacon pens, but it is much better to run them in well-strawed yards. Once they reach 50 kg they should be transferred to the fattening house and grouped in pens according to their size. If possible, keep hogs and gilts separate, as hogs grow slightly faster than gilts. This will mean that the pens of pigs will be ready for slaughter in more even batches.

Mixing strange pigs together often leads to fighting and ear biting. This can be prevented to some extent by feeding the pigs before they are mixed, and by spraying their backs with an aerosol scent which will destroy temporarily the pigs' own body odours. Empty paper meal sacks may be thrown into the pens for the pigs to play with. In the excitement of tearing the sacks

126

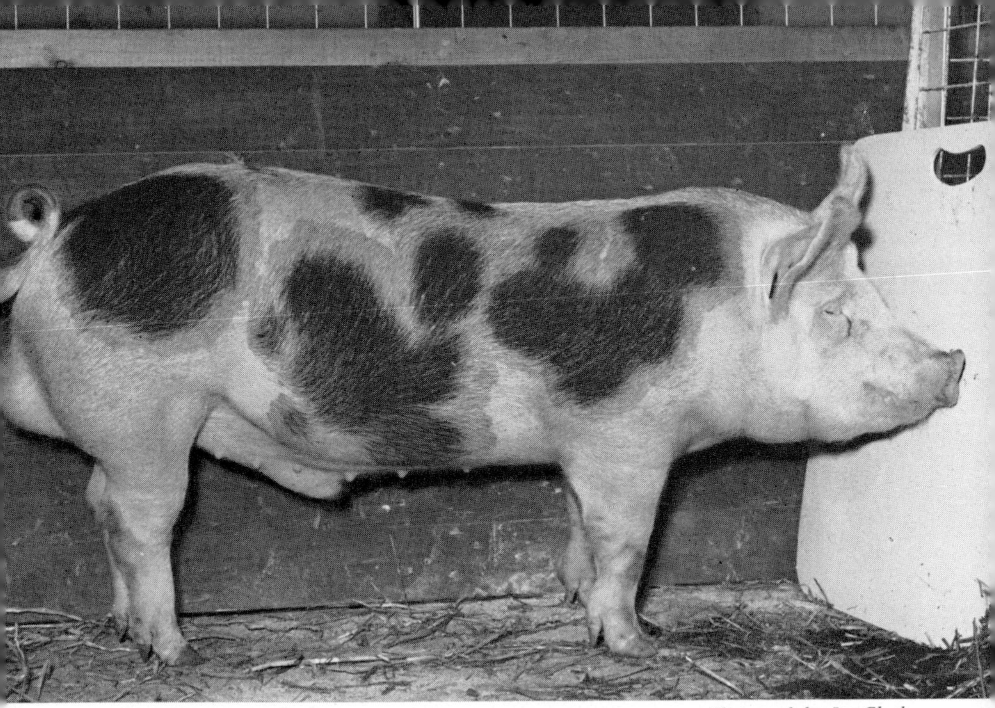

Photograph by Ian Clook

Pietrain boar

Hampshire boar

Photograph by Jack Fisher

Lacombe boar

Photograph by Jack Fisher

up, the pigs will rub against one another and in the process transfer their body odours.

You may also pour vegetable oil on the pigs' backs. Pigs love the taste of oil and will readily lick it off their new neighbours' backs. In a very short time the pigs will, generally, settle down.

Feeding

Once the bacon pigs are 50–55 kg and settled in the feeding house they should be carefully rationed according to their live-weights. Ideally, a pig should receive a different ration according to its age and liveweight, but today most farmers feed a high-protein sow and weaner ration to bacon pigs throughout their life, thus avoiding any check in growth rate that would result in changing the feed. Normally, we feed pigs twice per day, but *Dr. Braude* has shown in trials at the National Institute for Research in Dairying that bacon pigs will do equally well if fed their daily allowance once a day.

If once-a-day feeding is adopted, it is best to give the pigs their food in the morning, clean out the pens, inspect the pigs and then leave them undisturbed for the rest of the day.

Each bacon pig will require at least 225 mm of trough space when 50 kg liveweight, and this should be extended to 300 mm as the pig approaches slaughter weight, one method being to remove the odd pig from the pen. About 0·5 m² of floor space per head will be needed for lying down. Make sure that the pigs lie on a dry, comfortable floor, free from draughts and damp.

Bedding

You may bed the pigs down with long straw, chaff, bracken, peat moss, or sawdust and wood shavings. Straw is the most common litter, and where possible wheat straw should be used rather than barley straw, which tends to irritate the pigs' skin.

Peat moss is not very suitable because of its dusty nature, which tends to make the pigs sneeze. However, it should be remembered that peat moss is capable of absorbing eight times its own weight of moisture, and so may be an advantage in damp buildings or badly drained pens.

The pens and dunging passage should be cleaned out daily, and any damp or soiled bedding removed. Once a week thoroughly wash the dunging area down with water and apply a disinfectant to control disease and roundworm eggs.

Between each batch of pigs the pens should be thoroughly scrubbed with hot water and common washing soda to remove all traces of dung and to kill bacteria. After the pens have dried it is wise to whitewash the walls.

Fig. 45.   Washing pen

Health

It is important that every animal on the farm is inspected daily to see that it is fit and well. With pigs, you should note that they have an alert appearance, healthy appetite and come readily to the trough for their food. Examine the faeces to make sure that the pigs are not scouring or constipated. Listen for coughing which would lead you to suspect roundworms or enzootic pneumonia. Check carefully the pigs' skin and coat to see that it is not infected with lice or mange. Finally remember that a curly tail is an extremely good indication of a healthy pig.

Weighing

Weighing pigs regularly is an essential part of good management, for you will then know how your pigs are progressing, and have a measure of their rate of food conversion.

Weighing may be carried out either weekly or fortnightly, the latter being adopted by most producers. The pigs should be weighed at the same time each week, ideally in the early afternoon when the pig's stomach is almost empty; but since they may be restless at this time, many farmers choose to weigh in the morning after the pens have been cleaned out.

There is no need to identify each individual at weighing. All that is necessary is to record the total weight of the pen, and then find the average weight per pig.

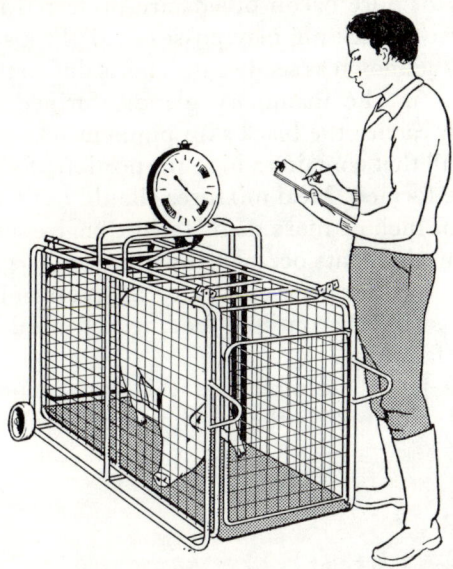

Fig. 46.   Weighing bacon pigs

e.g.

Pen of ten pigs weight	700 kg
the average weight per pig	70 kg
Previous week's weighing	660 kg
Average weight per pig	66 kg
∴ each pig has gained	4 kg

Food fed 2 kg meal/day × 7 days 14 kg meal

The food conversion is therefore $\dfrac{14 \text{ kg meal}}{4 \text{ kg liveweight gain}}$

We can see from this example that the pigs are doing reasonably well on this food, and have a conversion ratio of 3·5:1.

As the pigs approach bacon weight, they must be weighed weekly in order that they may be slaughtered at the right weight. The weight range is between 85–100 kg liveweight. It is best to enter your pig at 90 kg, since it may take 7–10 days before the factory will take delivery. The pigs should then dress out in the top bacon range of between 65–75 kg deadweight.

Strain

The final carcass will be influenced by the strain of the pigs, the foods fed, and, to some extent, the method of feeding.

White pigs of pure bacon breeds are preferred to coloured breeds, because a black pig may possess a slight discolouration in the belly meat known as seedy cut. This is due to the development of ducts in the mammary glands, formed by skin in growth, which carries the black skin pigment into the meat.

Feeding swill that contains a high proportion of oil will cause a soft fat to be formed. Meal mixtures should not contain more than 4% oil, or include more than 30–40% maize meal, linseed meal or rice meal. Taints occasionally occur in bacon, and this can be due to feeding poor-quality or an excess of fish meal, meat meal or swill. Discontinue feeding fish meal during the last month of fattening.

Hard white fat is desirable and this can be produced by feeding either skim milk, barley meal, tapioca meal, pea or bean meal.

Summary

Good strain of pig essential.

Cross-bred bacon breeds, for example Landrace × Large White is very suitable.

Worm weaners before mixing in matched lots.

Warm, dry, draught-free housing must be provided. Allow 300 mm trough space and 0·5 m² sleeping area per pig.

Feed only fresh, palatable high-quality sow and weaner meal from weaning to 45 kg—may be fed *ad lib*.

Restrict meal in final stages of fattening to 2·5 kg per head for cross-breeds, 2·7 kg per head for pure-bred bacon breeds.

Market at around 90–95 kg liveweight.

Dressing 73%–75%.

Pork production

Basically, the day-to-day management of pork pigs is the same as for bacon pigs. The weaners should be grouped in matched lots and housed in comfortable quarters. Feeding from self-feed hoppers is recommended to encourage maximum growth. Sow

130

and weaner meal is fed until slaughter, or a mixture of skim milk and barley meal. Cooked swill and potatoes are not suitable for porkers, as they tend to produce a poorer carcass.

Porkers are slaughtered at 45–55 kg liveweight, which will produce a carcass of around 32–40 kg. It should be remembered that pork tends to be a rather seasonal dish, being most popular in the winter months. Also, there is a wide variation throughout the country as to the type of porker required. Some butchers will buy well-fleshed flue-sheeted cross-breds and trim excess fat off the carcass; other butchers insist on long, thin, light-weight pigs from the bacon breeds, such as the pure-bred Landrace.

Before you decide to start up in pork production you would be wise to study your local markets to find the type of pig required.

Heavy-hog production

Heavy hogs are usually produced from cross-bred 'blue'-sheeted pigs such as the Walls Hybrid, or the Large White, Welsh or Landrace boar mated to Wessex, Essex, Gloucestershire Old Spot or Large Black sow. Blue pigs possess hybrid vigour, are capable of rapid growth, and produce well-fleshed carcasses, coupled with economic food conversion.

Feeding and management

Weaners should be first wormed, and then mixed in matched lots of up to fifty pigs per pen. Well-strawed yards with a small sleeping area make ideal accommodation. The weaners should receive good-quality sow and weaner meal *ad lib* from self-feeders. By-products such as swill or cooked potatoes may be used to replace part of the meal once the pigs reach 27–30 kg liveweight. If available, skim milk or whey may be fed, in addition to meal.

At approximately 60 kg liveweight the sow and weaner ration should be gradually changed to fattening mixture, with added minerals and vitamins. 90% barley meal, 8% soya-bean meal, plus 2% minerals will give satisfactory results, but it is better to mix two or three cereals together, such as:

131

60% Barley meal	50% Barley
20% Maize meal	40% Wheat (coarsely ground)
10% Wheat (coarsely ground)	8% Soya bean
8% Soya bean	2% Minerals
2% Minerals	

Minerals and vitamins are essential for good growth rate. There are several proprietary brands on the market.

Water must be available at all times when self-feeders are used. Water bowls or nipple drinkers are most suitable, provided they are sited away from the sleeping area.

Summary of heavy-hog production

1 Cross-bred pigs most suitable, for example Large White × British Saddleback.
2 Worm weaners, mix in evenly matched lots of 30–50 pigs.
3 Cheap housing, such as well-strawed yards, may be used.
4 Feed *ad lib*—sow and weaner, weaning—60 kg.
8% soya bean, 90% cereal and minerals   60–120 kg.
5 Swill, potatoes, skim milk, whey, biscuit waste and other by-products may be used to replace part of the meal allowance.
6 Heavy hogs should reach 120 kg by 6½–7 months.

Eighteen Pig Housing —
I Basic Principles

There is no one right way of raising pigs, or indeed housing them, but the basic principles of good housing are exactly the same whether we are using old adapted buildings, cheaply constructed straw-bale huts, or expensive controlled-environment houses.

To understand these principles the farmer must know something of the anatomy and physiology of the pig, a few elementary laws of physics, a knowledge of building materials, a thorough understanding of pig management, and an appreciation of the pigman's needs.

General considerations for the pig and pigman

The pig	The pigman
Warmth and fresh air	Amenable working conditions
Comfortable sleeping quarters	
Freedom from damp	Ease of watering and feeding
Adequate feeding room	Positive means of observing stock for signs of health and disease
Food and water supply	Labour-saving method of cleansing pens
Dung removal	Arrangements for weighing, handling and transporting stock

The pig is a warm-blooded animal; its normal body temperature is 38·9°C. It has no sweat glands. Baby piglets are poorly provided with hair (unlike calves and lambs), and so are particularly susceptible to cold and damp.

Pigs will always try to maintain their body temperature, no matter what the atmospheric temperature is. They do this by seeking the warmth of an infra-ray lamp or by burying themselves under loose dry straw or huddling up against their

133

neighbours. As pigs grow older, they develop resistance to cold by laying down fat under the skin. Adult pigs will in fact withstand a great deal of cold. Pigs exposed to the cold will require more food in order to maintain body temperature and to produce subcutaneous fat. Consequently, the food-conversion ratio will widen, and, of course, profit is affected.

Dampness is one of the worst enemies of young and old pigs alike. It leads to coughs, chills and enteritis in young pigs, and encourages rheumatism and leg weaknesses in older stock.

We must also remember that pigs have no sweat glands. This means that in very hot conditions, or when the atmosphere is still and humid, the pig will have difficulty in getting rid of body heat. It leads to rapid breathing, panting and more water is drunk. Given the opportunity, pigs will wallow in a mud bath or pools of urine in the dunging passage of a dirty pen.

Pigs need plenty of fresh air, especially when they feed, sleep and dung in the same pen. Fresh air will replace foul-air-carrying moisture, smells and disease-forming organisms. Good ventilation is a vital feature in pig housing. Closely related to ventilation is insulation, for unless the warmth that is produced by the pig's body is retained within the building, the pens will be cold and uncomfortable.

Pigs may be fed from troughs, self-feed hoppers or on the floor, but whichever system is adopted, you must see that each individual receives its full share.

A clean unfailing water supply is essential, for pigs cannot live for more than seventy-two hours without water. Water may be provided in either water bowls, automatic nozzles, or in troughs.

Remember also that the pig has a natural capacity for breaking out, and has no respect for rickety doors with broken hinges.

From this discussion we can see that when we house pigs we govern their environment, be it good or bad. If your house suits the pig's needs, then it will thrive, and you will prosper. If your housing is poor, the pig's performance will be likewise.

The pigman

Although it is sometimes argued that labour is only 10% of the costs in pig production, it should not be thought that the pigman's job is unimportant. Every effort should be made to help the pigman to do his job just a little bit better, for it is, in the long run, the interest and aptitude of the man in charge of the stock that makes the business a success or failure.

We should design pens in such a way that feeding may be done quickly and easily. We must minimise 'wasteful walking', e.g. carrying buckets of water, when a water bowl could be used, or carrying buckets of meal, when a barrow would save umpteen journeys. If it is possible in any way to lighten the task of 'mucking out', this should be done. Why not consider slurry cleaning, or mechanical cleaning, and try to dispense with the broom and shovel.

If more of the pigman's time is spent in looking at his pigs to make sure that they are fit and well, keeping accurate food and liveweight records by weighing the pigs regularly, rather than spending most of the day cleaning out, then he will be happier, better at his job, and consequently the business will become more profitable for the farmer. Weighing pigs is an important managerial task, but all too often neglected because of the lack of suitable handling facilities. When designing your new building be sure to incorporate a weighing pen where the scales are always available.

Lastly, let us not forget handling and loading pigs. Whether you market pork, bacon or heavy hogs, they must be loaded on to a trailer or lorry when they leave the farm. It is well worth while building a proper loading ramp for this purpose. It will save a great deal of time and loss of temper!

To summarise, we need to provide the pig with warm, dry, comfortable pens, free from draughts, but well ventilated. The pen should be easy to clean, and feeding arrangements should suit both the pig and pigman. A clean supply of water is most important. Handling arrangements for weighing and loading should be provided.

The site

The site for a new piggery should be chosen with care, for it will play an important part in determining the sucess of the new building.

Where possible, choose a near-level plot, on free-draining soil with an open aspect, preferably facing south. Avoid sites that are in frost pockets, on waterlogged soils, or exposed to strong winds. It will be advantageous if the site is near to a hard road, for today pigs and feedingstuffs are transported in increasingly larger vehicles, such as the bulk-feed tankers, and articulated lorries. Mains supplies of electricity and water should be available.

General construction

Materials

The shell or outer building may be constructed with either wood, brick, pre-cast concrete panels, galvanised iron or compressed asbestos. Table 14 shows the relative K values for various materials, which, together with price, should be considered before building commences.

Wood has long been popular, for it is the least expensive, and ideally suited for prefabricated factory-built piggeries, which can be quickly and easily erected on the farm. Wood is less durable than the other materials mentioned, but if it is creosoted regularly, it should last for at least 10–15 years.

Concrete blocks are popular with 'do-it-youself' farmers. They are strong, relatively easy to build and extremely durable. They make a compromise between the merits of brick and wood.

Brick is rather expensive for general use, but may be used for specialist houses, or where cost is not of prime importance.

Galvanised iron is rarely used in permanent buildings because of its poor insulating qualities. However, it is most suitable for outdoor range shelters, being exceptionally strong and long lasting.

Asbestos sheeting is generally used for roofing, but is unsuitable for walls, due to its brittle nature. However, recently 12 mm-thick compressed asbestos sheets have become available which are much stronger than the conventional sheet. They have the advantage of being easy to clean, and take up less room than brick or block walls.

Table 14. K Values of common building materials

Thermal conductivity (K)

The thermal conductivity of a material is a specific property of that material and is defined as the quantity of heat in watts per metre per degree Celsius (W/m deg C) The lower the K value of a material, the better its insulating efficiency.

Material	Density kg/m³	K value W/m deg C	Material	Density kg/m³	K value W/m deg C
Artificial aggregate	56–1750	0·173–0·577	Concrete—cont. no fines 1:10	1842–1922	0·634–0·937
Asbestos-cement sheet	1362	0·216–0·228	no fines 1:6	1442–1842	0·562–0·750
	1522	0·288	clinker	1522–1682	0·332–0·404
	2000	0·433–0·650			
			Fibreboard	224	0·052
Bitumen	1057	0·159		384	0·063
Bricks:			Glass	2500	1·05
common	1762	0·808			
engineer-	2195	0·793	Hardboard	560	0·079
ing	1041	0·317		753	0·094
lightweight	1362	0·375		881	0·123
				993	0·144
Brickwork, common	—	1·15	Mortar: cement:		
Building paper	—	0·065	sand 1:3	1890–2000	0·880–1·12
			cement:		
Coke breeze slab	—	0·577	sand 1:4	1954	0·923
			Compo-Portland		
Concrete:			cement:		
aerated	496	0·117	hydrated		
	800	0·173–0·202	lime:sand 1:3:9	1858	0·692
			lime-sand		
ballast			1:2	—	0·476
1:2:4	2243–2483	1·44			
			Pitch	—	0·144

Material	Density kg/m³	*K* value W/m deg C	Material	Density kg/m³	*K* value W/m deg C
Plaster:			Roofing felt	960	0·187
powder	993	0·144			
gypsum	1121	0·375	Slate	2723	1·87
perlite	400	0·079			
lime:sand:			Stone:		
cement	1442	0·476	artificial	1762	1·33
sand:			granite	2643	2·884
cement	1565	0·534	limestone	2178	1·53
sand:			marble	2723	2·549
gypsum	1410	0·650	sandstone	2000	1·30
vermiculite:					
gypsum	640	0·187	Thatch	240	0·072
vermiculite:					
cement	448–	0·108–	Tiles:		
	768	0·202	burnt clay	1922	0·836
			concrete	2162	1·15
Plasterboard	960	0·159			
			Timber*		
Plywood	528	0·138	(across		
			grain)		
Roofing			beech	705	0·167
asphalt	960	0·144	deal	609	0·125
	1600	0·433	oak	769	0·160
	1922	0·577	pine	657	0·138
			spruce	416	0·105

* 10 to 15 per cent moisture content

Table 14 is reproduced by special permission from the *Insulation Handbook*

Thermal Conductivity Values

K Values (Btu in/ft² h deg F to W/m deg C)

Btu	0·00	0·01	0·02	0·03	0·04	0·05	0·06	0·07	0·08	0·09
0·10	0·0144	0·0159	0·0173	0·0187	0·0202	0·0216	0·0231	0·0245	0·0260	0·0274
0·20	0·0288	0·0302	0·0317	0·0332	0·0346	0·0361	0·0375	0·0389	0·0404	0·0418
0·30	0·0433	0·0447	0·0461	0·0476	0·0490	0·0505	0·0519	0·0534	0·0548	0·0562
0·40	0·0577	0·0591	0·0606	0·0620	0·0634	0·0650	0·0663	0·0678	0·0692	0·0707
0·50	0·0721	0·0735	0·0750	0·0764	0·0779	0·0793	0·0808	0·0822	0·0836	0·0851
0·60	0·0865	0·0880	0·0894	0·0908	0·0923	0·0937	0·0952	0·0966	0·0981	0·0995
0·70	0·101	0·102	0·104	0·105	0·107	0·108	0·110	0·111	0·112	0·114
0·80	0·115	0·117	0·118	0·120	0·121	0·123	0·124	0·125	0·127	0·128
0·90	0·130	0·131	0·133	0·134	0·136	0·137	0·138	0·140	0·141	0·143
1·00	0·144	0·146	0·147	0·149	0·150	0·151	0·153	0·154	0·156	0·157
1·10	0·159	0·160	0·162	0·163	0·164	0·166	0·167	0·169	0·170	0·172
1·20	0·173	0·174	0·176	0·177	0·179	0·180	0·182	0·183	0·185	0·186
1·30	0·187	0·189	0·190	0·192	0·193	0·195	0·196	0·198	0·199	0·200
1·40	0·202	0·203	0·205	0·206	0·208	0·209	0·211	0·212	0·213	0·215
1·50	0·216	0·218	0·219	0·221	0·222	0·224	0·225	0·226	0·228	0·229

e.g. to convert 0·23 Btu in/ft² h deg F read across from 0·20 in left hand column and down from 0·03 in top line i.e. 0·0332 W/m² deg C

Reproduced by special permission from the *Insulation Handbook*

Table 15. Thermal transmittance—coefficient U-values of common building materials

U values or the 'thermal transmittance' coefficient means the number of Joules transmitted per second ($J/s = W$) through one square metre of the structure when there is a difference in the temperature of 1 degree Celsius between the air on either side of the structure. This is abbreviated as W/m^2 deg C.

The following table gives the U value of various constructions and materials used in pig housing. The lower the U value, the better will be the insulation.

	W/m² deg C
Walls	
Hardboard-polyurethane foam laminate 25 mm thick	0·68
Brick cavity wall (280 mm) plastered inside	1·7
Brick solid wall 230 mm thick	2·67
Stone, medium porous 300 mm	2·84
Tongue-and-grooved boarding 25 mm	2·84
Solid concrete 200 mm	3·18
Solid concrete 150 mm	3·58
Asbestos cement sheeting 6·0 mm	5·05
Corrugated asbestos sheets on steel framing	6·53
Corrugated iron	8·52
Roofs	
Corrugated asbestos—25 mm expanded polystyrene	0·88
Tiles or slates on boarding and felt with 50 mm strawboard	1·25
Corrugated asbestos with 300 mm loose straw	1·25
Corrugated asbestos—50 mm strawboard	1·25
Corrugated asbestos—50 mm glass fibre quilt	1·75
Corrugated asbestos—12 mm boarding	2·16
Corrugated asbestos sheets	7·95
Corrugated iron sheets	8·52

Ventilation

Ventilation is the removal of foul, moisture-laden air and replacing it with clean fresh air.

There is no doubt that pigs do better when they are housed in comfortable airy conditions, kept at the correct temperature. To achieve optimum conditions the insulation and ventilation must be correct.

140

Optimum temperature and humidity

	Relative humidity	°C
New-born piglets	70%	21–27
Weaners	70%	21–24
Pork	70%	18–24
Baconers	70%	16–21
Heavy hogs	70%	10–16
Farrowing sow	70%	16–21

Humidity is of extreme importance, for scientists have shown that for every 90 kg bacon pig in a house about one litre of water per day has to be removed from the building by the ventilation system. A relative humidity of 70% will provide an atmosphere which feels dry and will prevent condensation in a well-insulated building.

Ventilation system

There are three main systems of ventilation:
1 Natural ventilation
2 Forced ventilation
3 Pressurised ventilation

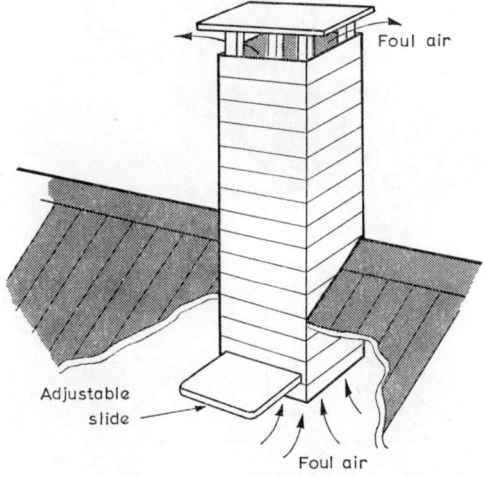

Fig. 47.   Air outlet

The first method is to extract air from a chimney-type construction fixed in the roof apex, and to allow fresh air into the building through hopper-type windows or baffled inlets.

141

The outlet area should be approximately 32 cm² per 45 kg live-weight or about 64 cm² per bacon pig (90 kg). Thus for every 100 bacon pigs housed you will need an extractor of 100×64 cm². This would be achieved with a ventilation shaft measuring approximately 80×80 cm.

Air inlets

The inlet area should be about three times the outlet area, thus we reckon approximately 100 cm² per 45 kg pig or 200 cm² per baconer. The inlets should be fixed in the side walls, at least 1 m above floor level, and not less than 0·3 m below the eaves.

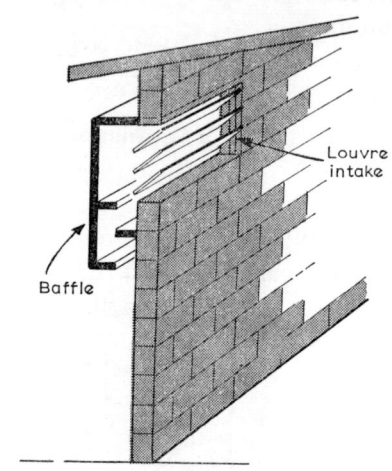

Fig. 48. Air inlet

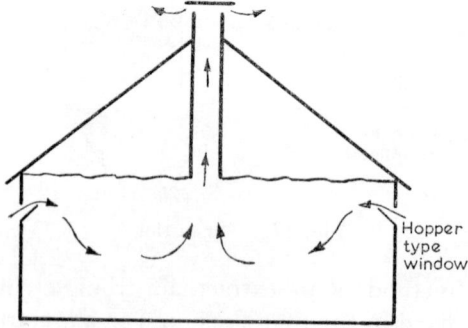

Fig. 49(a). Ventilation natural

142

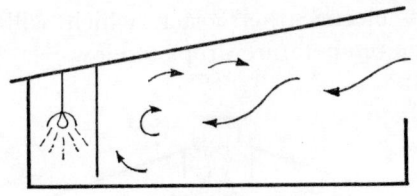

Fig. 49(b).   Ventilation natural

The disadvantages of natural ventilation are that the system cannot be controlled automatically, and therefore labour must be available to alter the inlets according to the outside weather conditions.

Forced ventilation

Forced ventilation entails the use of an electrically operated extractor fan fixed in such a position as to draw out foul air without causing draughts in the building. Usually the fans are fixed in or near the dunging passage, in order to extract the foul air from as near the source as possible. Fresh air is drawn in from a roof inlet.

A great deal of research has been carried out to determine the amount of air movement necessary to allow fresh conditions, and how this can be achieved with the use of different-sized fans, speed regulators and thermostats which will give a guaranteed controlled condition within the house.

Scientists have suggested that the pig requires a minimum of 0·3–0·38 cubic metres of air per hour (m³/h/kg) per kilogram liveweight during the cold winter months, and 0·8–2·0 m³/h/kg liveweight in the summer months. This means that in a building housing 100 bacon pigs the maximum air movement will be:

If 1 kg liveweight requires 1·0 m³/h
then a bacon pig (90 kg) will require 90 m³/h
therefore 100 bacon pigs ×90 m³ = 9000 m³/h

We must now purchase either a single fan capable of removing 9000 m³/h or say three smaller fans, each able to move 3000 m³/h.

One of the disadvantages of extractor fans is that during the winter months, when only a small air movement is required, the houses may suffer from a drop in temperature if the fans are run fully. To overcome this problem it is necessary to connect

143

the fan to an electric thermostat, which will stop the fan operating if the temperature drops too low.

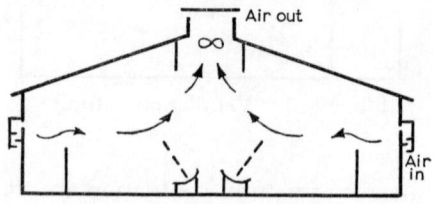

Fig. 50(a).   Ventilation forced

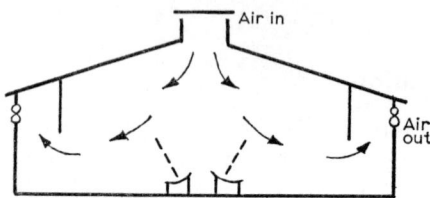

Fig. 50(b).   Ventilation forced

Pressurised systems

A recent development in ventilation is to draw fresh air into the piggery through a central position in the ceiling by means of an impeller fan. The air within the building will become pressurised, and therefore as the pressure increases, foul air will be forced out through side vents. The main advantage of this system is that incoming draughts are virtually excluded because of the air pressure within the building. The impeller fan is connected to a thermostat so that the inside temperature is easily regulated.

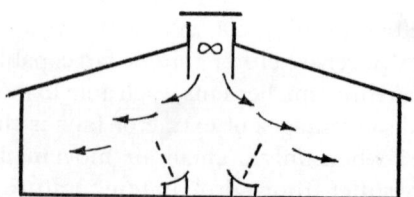

Fig. 51(a).   Ventilation pressurised

144

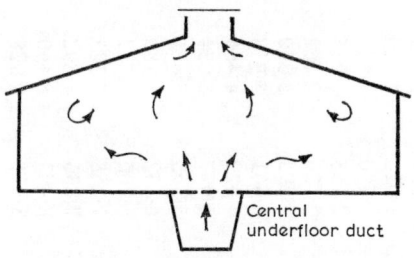

Central
underfloor duct

Fig. 51(b).   Ventilation pressurised

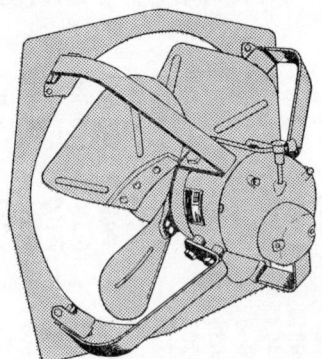

Fig. 51(c).   Propeller fan

Conclusion

It must be remembered that whichever system is used, the aim should be to provide warm, airy, draught-free conditions, with a low humidity. When you have fed your pigs always check that they are lying down comfortably in the sleeping quarters. Check that there are no draughts or strong smells inside the building.

Insulation

Insulation of the roof, walls and floor is necessary in order that the heat produced by the pigs' bodies is conserved within the building. It has been estimated that ten pigs of 90 kg liveweight will produce as much heat as a 2 kW electric fire.

It is generally agreed that a constant temperature of between 16–21°C with a dry atmosphere is preferable for fattening pigs. Higher temperatures lead to high humidity, if badly ventilated.

145

Thermal Conductivity Values

U Values (Btu/ft² deg F to W/m² deg C)

Btu	0·00	0·01	0·02	0·03	0·04	0·05	0·06	0·07	0·08	0·09
0	0·000	0·057	0·114	0·170	0·227	0·284	0·341	0·397	0·454	0·511
0·1	0·568	0·625	0·681	0·738	0·795	0·852	0·909	0·965	1·022	1·079
0·2	1·136	1·192	1·249	1·306	1·363	1·420	1·476	1·533	1·590	1·647
0·3	1·703	1·760	1·817	1·873	1·931	1·987	2·044	2·101	2·158	2·215
0·4	2·271	2·328	2·385	2·442	2·498	2·555	2·612	2·669	2·726	2·782
0·5	2·839	2·896	2·953	3·009	3·066	3·123	3·180	3·237	3·293	3·350
0·6	3·407	3·464	3·521	3·577	3·634	3·691	3·748	3·804	3·861	3·918
0·7	3·975	4·032	4·088	4·145	4·202	4·259	4·315	4·372	4·429	4·486
0·8	4·543	4·599	4·656	4·713	4·770	4·827	4·883	4·940	4·997	5·054
0·9	5·110	5·167	5·224	5·281	5·338	5·394	5·451	5·508	5·565	5·621
1·0	5·678	5·735	5·792	5·848	5·905	5·962	6·019	6·076	6·132	6·189
1·1	6·246	6·303	6·359	6·416	6·473	6·530	6·586	6·643	6·700	6·757
1·2	6·814	6·870	6·927	6·984	7·041	7·100	7·154	7·211	7·268	7·325
1·3	7·382	7·438	7·495	7·552	7·609	7·665	7·722	7·779	7·836	7·892
1·4	7·950	8·006	8·063	8·120	8·176	8·233	8·290	8·347	8·403	8·460
1·5	8·517	8·574	8·631	8·687	8·744	8·801	8·858	8·914	8·971	9·028

e.g. to convert 0·32 Btu/ft² deg F read across from 0·30 in left hand column and down from 0·02 in top line i.e. 1·817 W/m² deg C.

Reproduced by special permission from the *Insulation Handbook*

146

There is, however, a notable exception to this rule—*Mr. H. Jordan* of Northern Ireland keeps pigs successfully in high-temperature, high-humidity houses known as sweat boxes.

The following diagram shows how the heat produced inside an uninsulated building is lost.

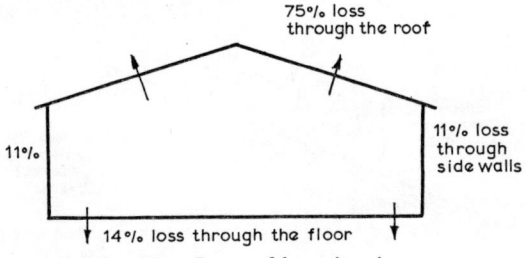

75% loss
through the roof

11% loss
through
side walls

11%

14% loss through the floor

Fig. 52. Loss of heat in piggery

U values

Insulation is usually achieved by means of a material that provides a 'static air space', for example loose straw, glass wool or expanded polystyrene. The insulating value of a material is measured by its thermal conductivity value. This is defined as *'the number of joules passing through one square metre of a material per second when there is a 1°C temperature differentiation on either side'* (W/m^2 deg C).

The thermal conductivity values are given in Table 15. It should be noted that materials with a value of 3·0 or more are not to be recommended.

Roof insulation

The function of the roof is to keep the wet out and the warmth

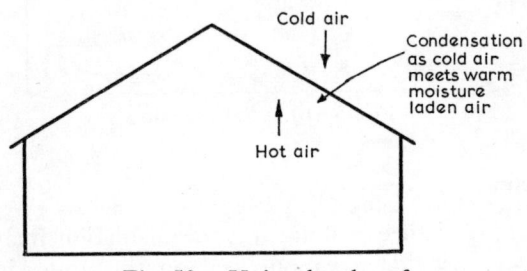

Cold air

Condensation
as cold air
meets warm
moisture
laden air

Hot air

Fig. 53.   Uninsulated roof

inside the building. In uninsulated buildings 75% or more of the heat produced may be lost through the roof. This is due to the warm air rising inside the house, and then being rapidly cooled as it meets the cold surface of the uninsulated roof.

If we place a non-conducting (insulator) material under the roof, the warm air will be kept inside the building.

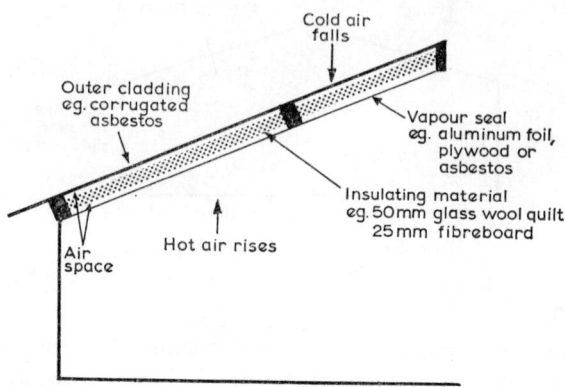

Fig. 54(a)   Roof insulation

In buildings with a high-pitch roof, or where an existing building with high eaves is being converted, it is better to construct a false ceiling about 2 m above the pigs.

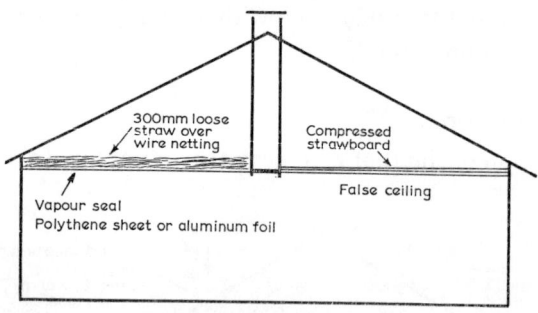

Fig. 54(b).   False ceiling

Vapour seal

To prevent moisture penetrating the insulation material it is necessary to provide a vapour seal on the underside of the

148

insulating material. Suitable vapour seals are asbestos sheets, hardboard coated with good-quality paint, aluminium sheeting, or the proprietary insulating boards may be purchased that have been specially treated with a sealing compound.

Floors

'A warm bed is worth an extra feed.' Fattening pigs spend about 80% of their time lying on the floor. We should take care, therefore, to ensure that floors are dry, warm and comfortable. Concrete is usually used because of its hard-wearing nature, but it tends to be cold unless a damp course and underfloor insulation is provided.

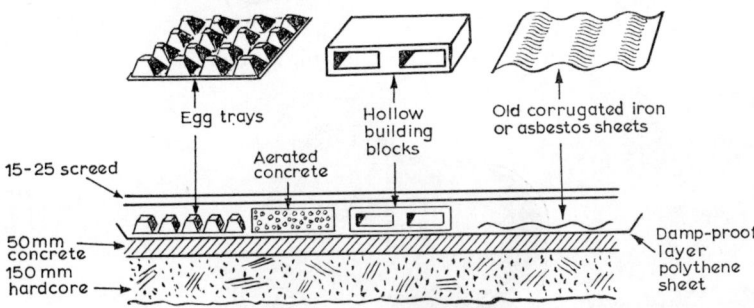

Fig. 55.   Floor insulation

Walls

Exterior walls are usually built with hollow concrete blocks, brick or timber. Ideally, a cavity wall should be built to give a high degree of insulation, but the cost of erecting double walls may be prohibitive.

Inside partition walls should be at least 1 m high for bacon pigs, and 1·2 m high for sows. They must be strong and easy to keep clean. Suitable materials are 25 mm thick timber boards treated with creosote to prevent the pigs gnawing holes, 100 mm concrete blocks, 100 mm bricks, 25 mm compressed asbestos or 50 mm pre-cast concrete slabs, the latter two materials having the advantage of taking less room, and may be built in such a way as to provide a movable partition.

149

Troughs, water bowls and nozzle drinkers

Feeding troughs may be of a portable or permanent design. Allow 250 mm of trough space for pork pigs, 300 mm for baconers, 350 mm for heavy hogs and 350–450 mm for sows.

Permanent troughs may be built with concrete, and lined with 250 mm wide glazed fireclay tiles, to give a smooth, hard-wearing, hygienic surface. Portable troughs are usually made with galvanised iron, although there are a few cast-iron troughs to be found on some farms.

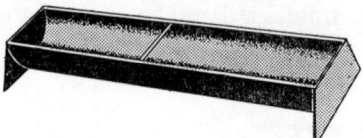

Fig. 56(a).   Portable pig trough

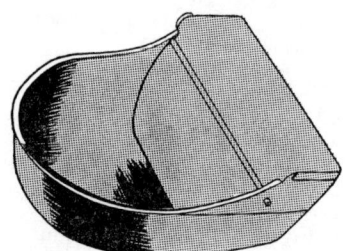

Fig. 56(b).   Water bowl

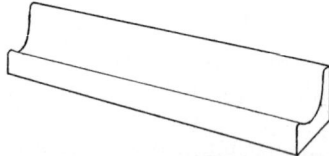

Fig. 56(c).   Glazed-tile trough

If floor feeding is adopted, then it will be necessary to provide water bowls or nozzle drinkers. These are best sited in the dunging area, as far away from the sleeping quarters as possible. This will encourage the pigs to use the dunging passage, and should a water bowl overflow, will not wet the pigs' bedding.

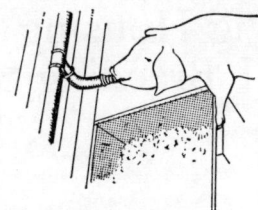

Fig. 56(d).   Pig drinker

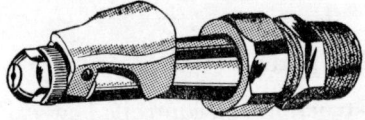

Fig. 56(e).   Nozzle drinker

Doors

Doors are the weakest part of the building, and so it is vital that they should be soundly constructed, well hung and secured in position with strong fasteners. Where wooden doors are used in the dunging passage they are best covered with sheet metal to protect the wood against wet dung and urine. Alternatively, 12 mm thick compressed asbestos sheets may be used.

Nineteen Pig Housing — II Breeding Stock

Indoor farrowing pens

The general construction of a controlled environment farrowing house is basically similar to the bacon house. The internal layout, however, is somewhat different and should contain a soundly constructed creep to provide warmth, food and protection for the piglets; a comfortable sleeping area for the sow; and a dunging and feeding area where a clean supply of water is always available. The partition walls should be at least 1 m high, but preferably built to ceiling height, as this will isolate the pens and help to prevent the spread of airborne disease. Either a farrowing crate or farrowing rails should be incorporated to help reduce the risk of crushing.

The creep is an essential part of any farrowing accommodation. Remember, the sow is a clumsy animal (particularly the white breeds), and we must, therefore, try to prevent her lying on baby pigs and killing them by crushing. The idea of a creep is to provide a small pen in which an infra-ray lamp is fitted. The lamp will raise the temperature of the creep to around 21–27°C, and the light will attract the newly born pigs away from the sow. The piglets return to the sow only when they are hungry.

Construction

Where possible, build the creep adjoining the feeding passage, so that the pigman has access for feeding and inspection without entering the sow's pen, allowing at least 0·13 m² per pig. Build the creep square if possible, 1·2 m × 1·2 m are satisfactory measurements. The square, rather than the oblong creep, will encourage a more even temperature inside the pen. The side walls must be strong. Solid walls are preferred to rails. A roof

152

over the creep is essential, for this will keep the heat in and help to prevent draughts at ground level.

Fig. 57.   Farrowing pen and creep

Farrowing crates

There are numerous makes and designs of farrowing crates on the market. Some are portable, while others are semi-portable or permanent. However, they all have the same objective, that of rearing more pigs per litter. The sow is restricted in the crate, and if this is well designed she will only be able to lie down by first kneeling with her front legs and then rolling on to her side. This gives the piglets sufficient time to move safely away, and if an infra-ray lamp is placed by the side of the crate the youngsters will be tempted to lie in the warmth away from the sow.

Although it was thought originally that sows should stay in the crates for about seven days (danger period), we find today that sows can be kept in the crates for up to three weeks without coming to any harm. This has now led to the development of controlled environment 'nurseries'.

Farrowing crate house (nurseries)

A fairly recent system which has been adopted on many farms is to farrow sows down in batches, using farrowing crates standing side by side in a controlled environment house. The pigs stay in the building for two to three weeks and are then transferred to cheaply constructed follow-on pens or the piglets are 'early weaned'. The crates are removed from the house for thorough cleansing and sterilisation.

This type of accommodation is less expensive than traditional pens, because only a low plywood wall is necessary between the crates. The risk of disease, however, is much greater, and it is, therefore, essential that the house is emptied, cleansed and rested between batch farrowings.

Fig. 58(a).   Farrowing crate house

Farrowing and rearing pens

On most general farms one pen or loose box is provided for the sow to farrow in and where her litter will remain for the next eight or ten weeks before they are taken to a weaner pool. In such a house provision must be made for protecting the baby piglets at birth and for the next few days, and also a creep should be provided for feeding the piglets supplementary food.

Farrowing rails may be placed around the walls—0·2 m high and 0·2 m from the wall to provide protection, or a swinging gate may be used to make a removable 'crate'.

The total pen size should be about 3 m × 3 m. This will allow sufficient room for a sleeping area of 4·5 m², a creep area of 1·5–2 m² and a dunging area of about 2 m².

154

Multiple suckling pens

Another fairly recent development in sow management is to mix three to five sows and litters together in one pen when the piglets are about three weeks old, the main advantage being that the piglets form a 'weaner pool' at an early age and the problems

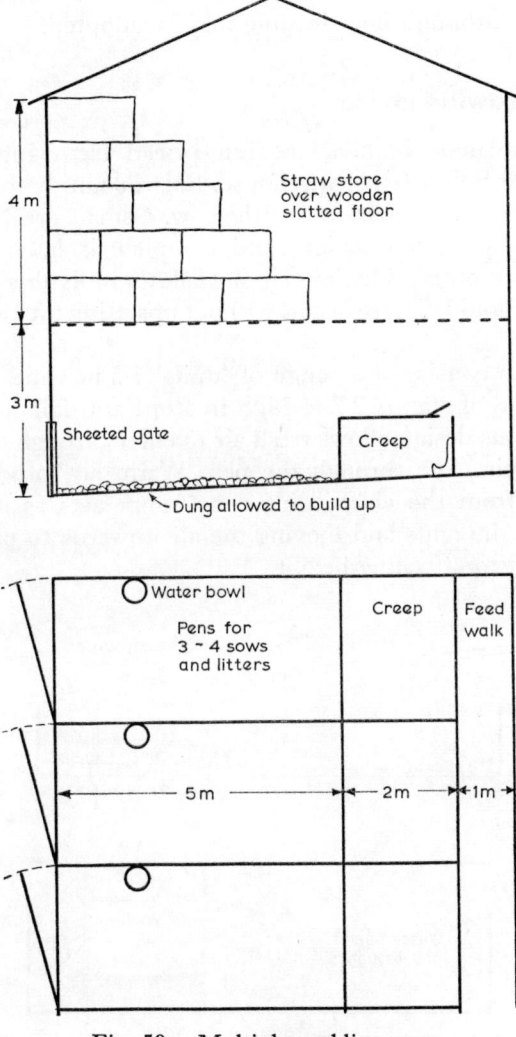

Fig. 59.   Multiple suckling pens

of 'stress' are kept to a minimum. Thus it is possible to house upwards of fifty pigs in one pool and keep them together until they reach about 40–50 kg liveweight.

There are several ways of multiple suckling. You may adapt existing building or erect purpose-designed structures, but the essential features are that each sow and litter has at least 4–5 m² of bedded area and a further 1·5–2 m² of the creep. Usually the pens are bedded with straw. Individual feeders for the sows are preferable, although floor feeding may be adopted.

Solari farrowing house

One of the most popular medium-priced farrowing houses today is the Solari farrowing house. Solari's aim is to provide warm, comfortable quarters for the sow, a small creep for the piglets, adequate ventilation, and complete isolation of one litter from another. The latter point allows individual pens to be rested, should disease occur, without upsetting the remaining pens.

The pens consist of a range of 'units' 1·5 m wide and 5 m deep. The roof stands 2·2 m high in front and falls to 1 m at the rear. This design allows fresh air to enter through the open front and circulate through the pen. Warm air, produced in the creep from the electric infra-ray lamp, acts as a buffer, preventing draughts and moving the air upwards to pass back out of the pen at ceiling height.

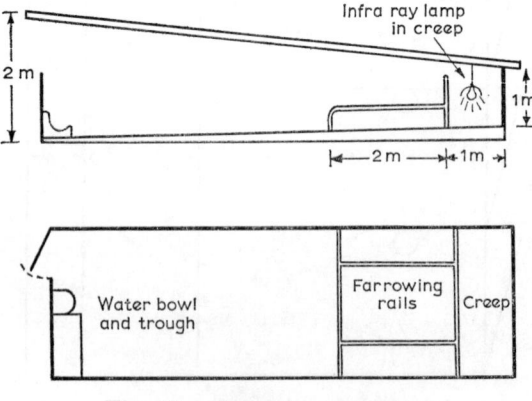

Fig. 60.   Solari farrowing pen

156

To be successful, the houses must face south, and during extremely severe weather hessian sacks or hardboard flaps may with advantage be suspended over the open fronts.

The only disadvantage to this house is the need for the pig-man to feed and muck out the pens from outside, and to walk past the sow to feed the pigs in the creep. This can be unpleasant with a nasty-tempered sow.

Dry sow accommodation

Traditionally the in-pig sow has been kept outdoors and brought inside for farrowing, but in recent years we have seen the rapid development of sow yards and more recently the sow stall.

Sow yards

Sow yards allow all the advantages of an outdoor system, fresh air, sunshine, exercise, etc., and have none of the disadvantages: rain, mud, laborious feeding and watering, fences to maintain, etc. They are extremely popular on arable farms as a means of treading straw into F.Y.M.

In an ideal sow yard the sows have a low-roofed sleeping area, and individual sow feeders, placed on a concrete apron on the opposite site of an open yard.

The pigman is able to observe the sows walking to and from the feeders, and he can feed the sows without entering the yard.

Where individual feeders are not available the sows should be fed with the large-size 'Jumbo' pig nut, scattered on top of the straw.

Dry-sow stalls

Sow stalls are the most controversial talking points in pig keeping today, praised by many scientists and farmers, yet condemned by the Brambell Committee.

The idea of keeping pregnant sows in stalls, which are smaller than farrowing crates and only allow the sow to stand up and lie down, was introduced a few years ago into this country from Scandinavia. The main advantages of the stall are that they provide low-cost accommodation, the sow needs less food,

157

there can be no bullying, and (it is claimed) more pigs born alive than with traditional methods.

Fig. 61.   Sow stalls

Management

The sows are introduced to the stall directly they are weaned, and confined in the stall until two or three days before their next farrowing. The only time they are removed is for mating with the boar.

Because the sows get no exercise they require less carbohydrate food, and up to 25% saving has been effected. It is, however, important to see that each sow receives adequate protein, vitamins and minerals. They must be kept warm, and even in the best houses some form of supplementary heating may be necessary in extremely cold weather. Good stockmanship is vitally important, for a careful eye must be kept on the sow's health, and especially on the condition of her feet and legs to see that no weakness occurs. Needless to say, the pens must be cleaned out daily and the floor area kept dry.

Tethered stalls

A further development of the sow-stall principle is tethered stalls. These differ from the stall house in that the stalls are shorter, about 1·5 m compared with 2 m, and the metal divisions are only 1 m long. With this arrangement it is possible to feed the sows from the rear, thus using the dunging area as a

158

feed walk. The great advantage here is that if the sows are tethered back to back considerable floor space is saved by eliminating the normal feed walk. A standard 9 m wide building will hold four rows of sows with this method.

Fig. 62(a).   Tether house

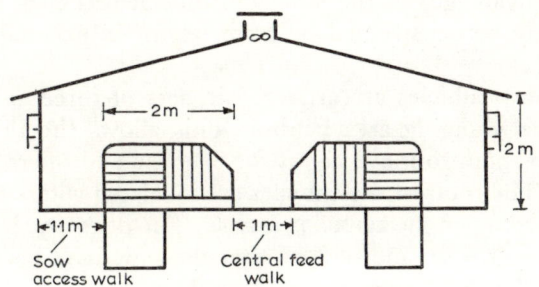

Fig. 62(b).   Sow stall

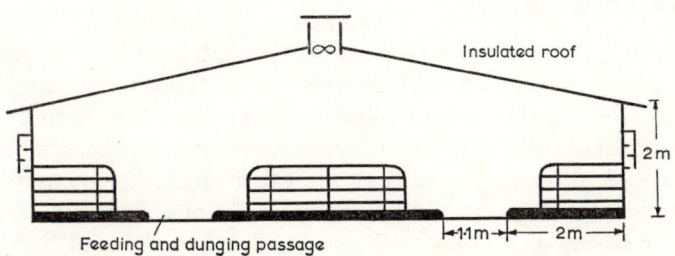

Fig. 63(a).   Tether house

159

Fig. 63(b). Tether house

Sow cubicles

Developed by Bill Marshall of B.O.C.M., sow cubicles combine all the advantages of the sow stall, but at less cost, and very often may be constructed by farm labour in existing covered yards.

The sow cubicles are arranged in sets of three and have a communal dunging area behind. This allows the three sows adequate room to move about for exercise and offers fresh air as well. The roofs of the cubicles are insulated with straw bales over boarding on the kennel principle. The pens may be cleaned out using a tractor and scraper after the sows have been shut in the cubicles by closing the partition gates.

Boar accommodation

Working boars may be either run with batches of in-pig sows, up to the third month of pregnancy, or housed separately in pens close to the sows. Probably the best method is to build strong boar pens with open yards where the boar can exercise and look through the rails to see other stock. On no account should a boar be kept in cold, damp conditions, right away from other stock, or he may well develop leg troubles and become vicious.

160

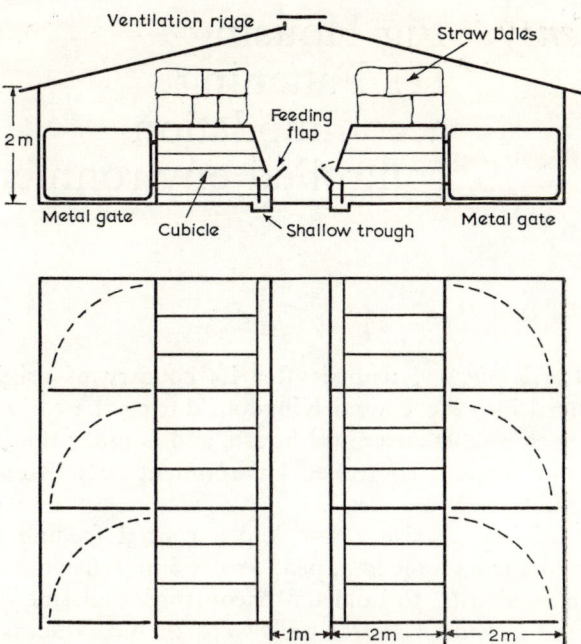

Fig. 63(c). Cubicle

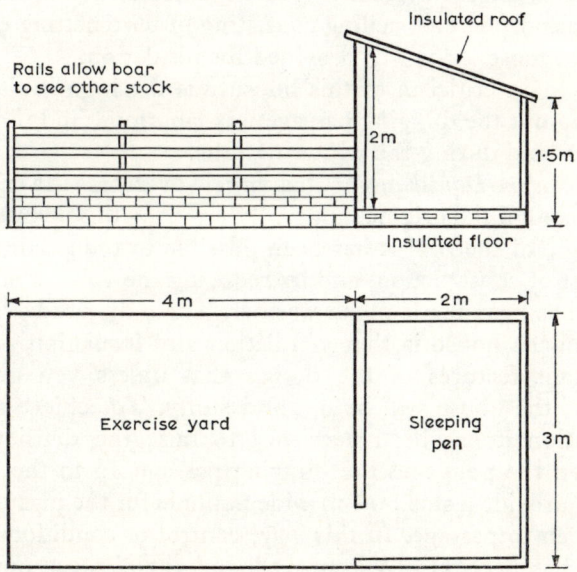

Fig. 64. Boar pen

Twenty

Pig Housing —
III Fattening
accommodation—
controlled environment

The Danish piggery, named after its country of origin, was
introduced into the United Kingdom during the early 1930's.
It has been a most successful house, and is today the basis on
which our modern controlled environment bacon houses are
designed.

The original Danish house had a central feeding passage
1·3 m wide, trough feeding, pens 3 m × 3 m (much too deep by
modern standards) to hold ten bacon pigs, and side dunging
passages 1 m wide. Exterior walls were 2·5 m to the eaves, and
supported a high pitch roof. This led to a large volume of air
in the building, which often created difficulty in maintaining a
warm atmosphere. A ceiling consisting of wire netting carrying
0·3 m of loose straw was provided for insulation.

The main criticism of this house was the high capital cost,
cleaning out the dunging passage was laborious, and the building was cold during the winter months.

The modern Danish piggery has largely overcome the criticisms
mentioned of the original house. By lowering the side walls and
the angle of the roof it has been possible to reduce drastically
the cost of construction, and by reducing the volume of air the
house has been made much warmer. The only disadvantage of
the modern house is that ventilation and insulation are such
important features of the design that unless you get both
correct, the house will be hot and stuffy. To achieve suitable
ventilation it has been necessary to raise the division walls
between the pens and the dunging passage up to the ceiling,
and to provide a small 0·4 m wide pophole for the pigs to move
from pen to passage. In this way, control of conditions in the
house has been greatly improved, and the pigs can now sleep
and feed in a warm inside shell. The outer dunging passage

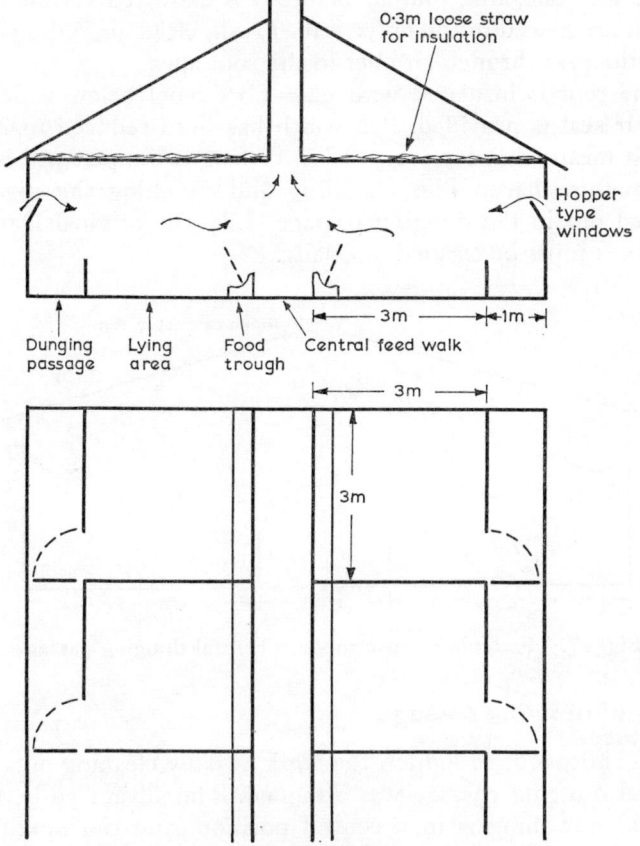

Fig. 65. Danish fattening house (1930)

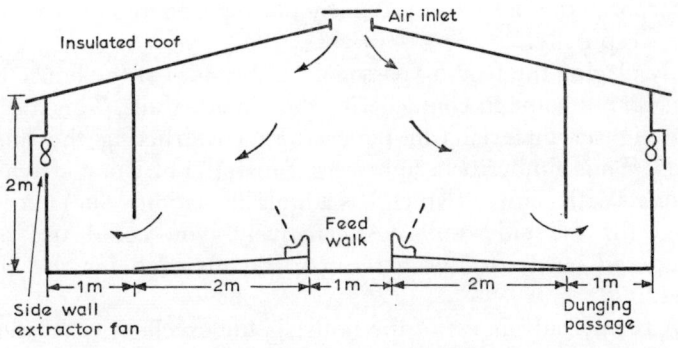

Fig. 66. Modern Danish

holds the stale, wet, foul air before it is extracted via the fans which are placed in the outer wall. Fresh, clean air is drawn in over the pigs through an inlet in the roof apex.

The roof is insulated with glass-fibre wool below which a vapour seal is provided. Pen width has been reduced to 2 m, which means that for every 0·3 m of length the pen will hold one mature bacon pig. Handling and weighing the pigs is carried out in the dunging passage. Like the original Danish house, it must be cleaned out daily.

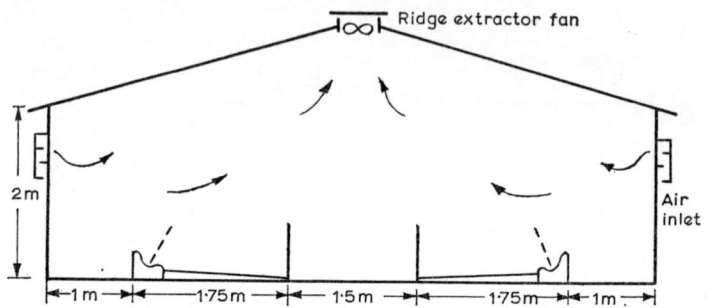

Fig. 67. Fattening house modern central dunging passage

Central dunging passage

In an endeavour to lighten the work of daily cleaning out, the central dunging passage was designed. The advantage here is that all the dung is in a central position, and can be either collected in a barrow or pushed out mechanically by a small tractor and squeegee. However, the central passage was later incorporated in a slurry system by placing a slatted floor over a 1 m deep gully.

By placing the feeding passages on the sides of the house the pigs do not come in contact with the outside walls. This means that lighter material can be used in constructing the outer shell. Thus timber or a lightweight material of wood shavings bonded with cement (which has a high insulation value) may be used for the side walls. Alternatively, you could use oil-tempered hardboard or exterior-grade plywood for the side cladding.

A further advantage of the house is the excellent ventilation. We know that hot air rises and cold air falls. It is, therefore,

164

quite easy to adopt natural-flow ventilation in a house with a central dunging area. The cold air comes into the building through ventilators in the side walls, passes over the pigs, and then, as it warms, rises up over the dunging passage, drawing out the foul fumes.

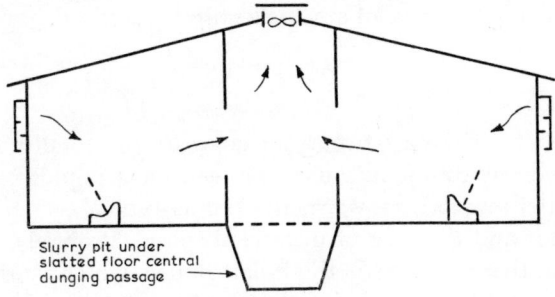

Fig. 68(a). Modern central dunging passage with slurry disposal

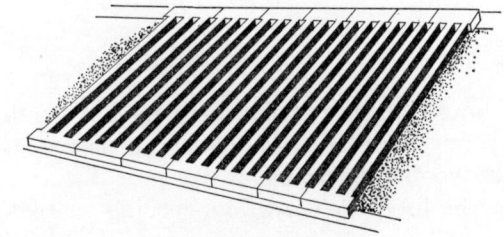

Fig. 68(b).   Slats

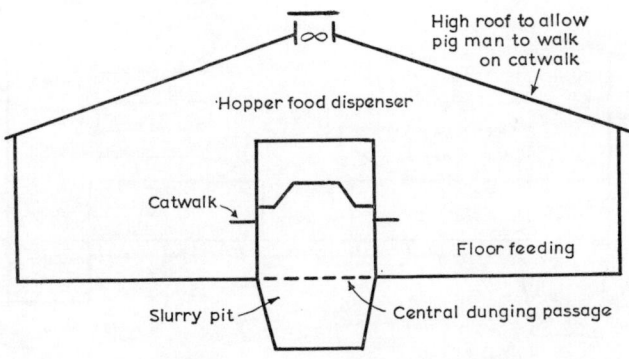

Fig. 69. Automatic feeding and cleaning

Perhaps the ultimate in controlled environment piggery design is the house with a central slatted floor dunging passage, coupled with overhead automatic feeding. The pigman is not required to clean or feed the pigs manually. He is an executive who is responsible for checking the system, and keeping a watchful eye for early signs of disease or vices occurring in the pigs. It must be emphasised that with these labour-saving systems a high degree of stockmanship is required.

Kennels

Although for the past twenty or so years the totally enclosed controlled environment house has been most popular, there is today a definite trend towards the kennel-type house due to the lower cost and the ease of manure disposal. Probably the best known of these designs is the Solari house, which is especially suited to arable farms, where very often an existing covered cattle yard or similar barn can be converted and where plenty of straw is available.

Solari kennel

The Solari house consists of sleeping pens built under an open barn. The kennel walls are 1 m high and pens are covered with 25 mm thick wooden boarding. A flap near the central feeding passage may be lifted to allow floor-feeding ventilation and to

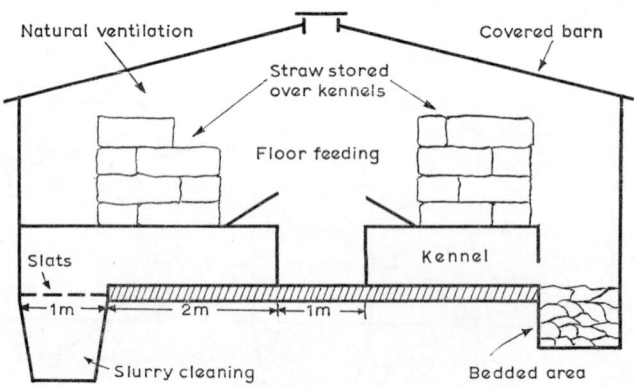

Fig. 70. Solari fattening house

inspect the pigs. The kennels may be further insulated by stacking straw bales on the boarded area. An open dunging area is provided which is bedded with straw and the muck allowed to build up. This need only be cleaned out when the dung reaches about 0·5 m in depth. Alternatively a slatted-floor slurry system may be used or daily cleaning may be adopted with a tractor and scraper.

Suffolk kennel

This is a very popular feeding house and may be constructed in two ways, either with simple kennels under an open dutch barn or secondly with purpose-designed fully insulated lying areas and a covered area over the feeding and dunging pens.

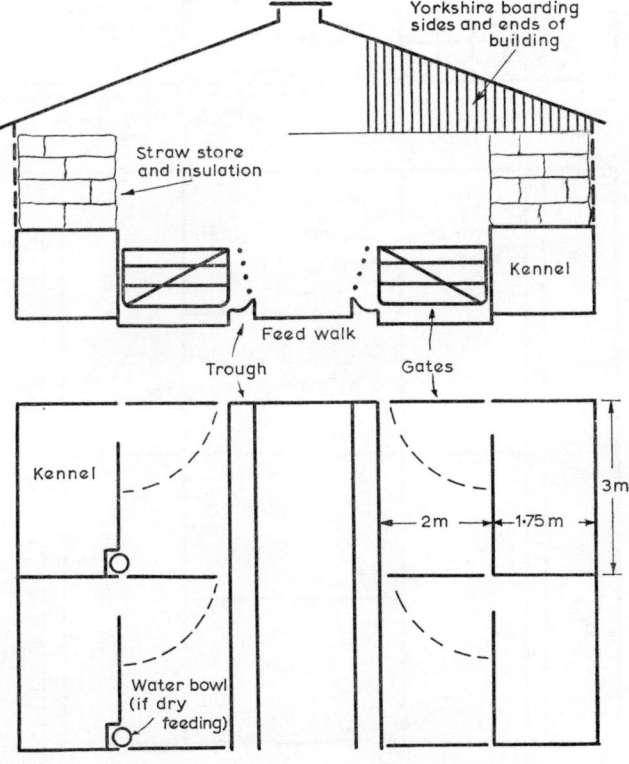

Fig. 71(a). Suffolk—purpose designed

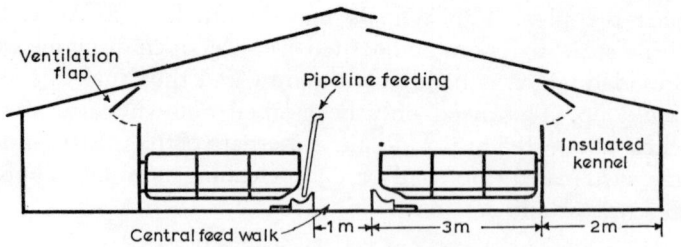

Fig. 71(b). Suffolk house

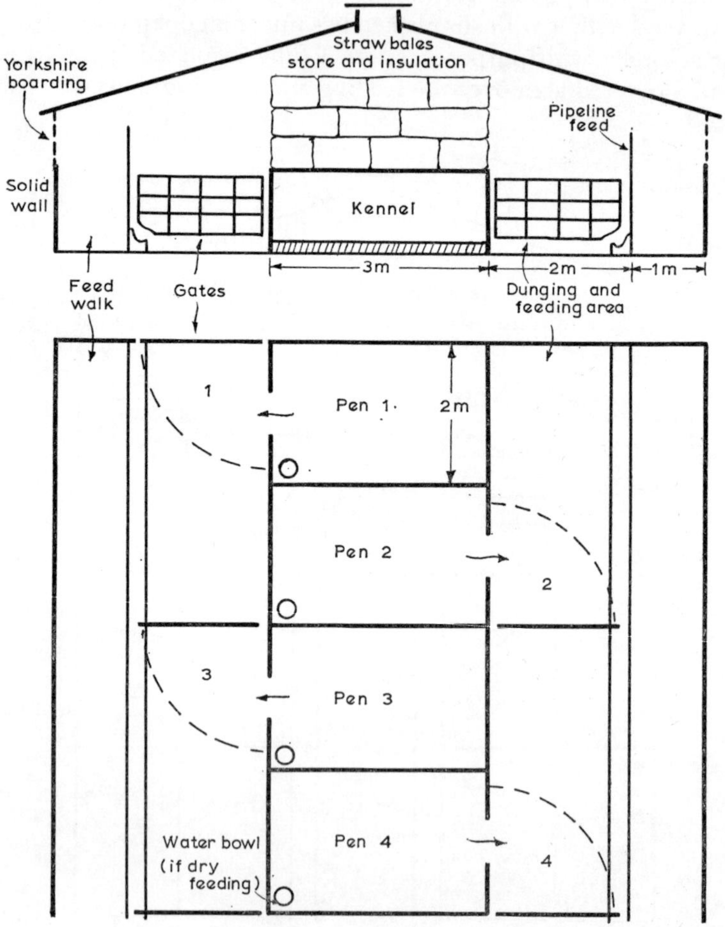

Fig. 72. Zigzag

168

The first type, see Fig. 71(a), is the cheapest method where an existing barn is available, and allows straw to be stored over the kennel. Feeding is usually by a pipeline and daily cleaning out with a tractor and scraper.

The second type (see Fig. 71(b)) is more expensive due to the cost of the insulated lying area. Ventilation is by means of an adjustable flap which is opened in hot weather and almost closed in the colder winter months.

Each pen is approximately 3 m wide and 1·5 m deep. This allows ten baconers per pen.

The outer barn is usually clad with 'Yorkshire boarding', which means spacing 100 mm × 25 mm boards, 25 mm apart. This will prevent strong winds blowing over the pigs, yet allow adequate ventilation.

Zigzag

This is basically the same as the Suffolk house except that the kennels are placed in the centre of the barn, which allows a greater storage space for straw. Because the sleeping pens are close together, heat loss is reduced. The disadvantage of this design is that two feeding passages are needed instead of one.

Twenty-one Pig Administration

The National Pig Breeders' Association

Pedigree pig breeding in an organised way started in this country in the late nineteenth century and led to the formation of the National Pig Breeders' Association (N.P.B.A.) in 1884, the principle objective of the association being 'maintaining the purity and improving the breed of swine in the United Kingdom of Great Britain and Ireland'. Since its formation, the N.P.B.A. has kept meticulous records of pedigree pig ancestry and publish each year a Herd Book for each breed in the association. It also arranged carcass competition and sales of pedigree pigs, organises pig classes at the major agricultural shows, and assists breeders who wish to export pedigree pigs.

Over the years the association has played a major role in pig improvement and was responsible for starting National Progeny Testing, which it later handed over to the Pig Husbandry Development Authority.

British Landrace Pig Society

The British Landrace Pig Society was formed in 1953 in order to maintain and improve the quality of the Landrace breed, which at that time had been introduced into the U.K. from Sweden (see page 43).

The Society established its own Herd Book to record all pedigree data for Landrace pigs bred in Great Britain and Northern Ireland.

The Society also organises shows and sales, publishes breed literature, makes arrangements for exporting Landrace pigs and generally furthers the interests of the breed.

FORM

DAM

Ear Number _____ H.B. Number _____ Breed _____

SIRE

Ear Number _____ H.B. Number _____ Breed _____

Date of Farrowing DAY | MONTH | YEAR

Number Born ALIVE DEAD

Birth Weight

Tick here if:-
(1) Artificial Service _____
(2) Got by Hysterectomy _____

(in Ear number space below)

Ear Number	Sex	Number of Teats	Weight 3 weeks	Weight 8 weeks	In/Out	Sow	Age at Slaughter (days)	Cold Weight (lb)	Length	Loin	Shoulder	Grading	Remarks, e.g. Genetic Defects or Date of Death and Cause
1													
2													
3													
4													
5													
6													
7													
8													
9													
10													
11													
12													
13													
14													
15													

(Fostered columns: In/Out, Sow)

	Ear Number	Date Administered	Nature of Medication

	Ear Number	Date Administered	Nature of Medication

Number alive _____
Total weight _____
Average _____

I hereby declare that to the best of my knowledge and belief the particulars entered are correct in respect of this litter.

Signed. _____

Date _____

STAMP NAME AND ADDRESS HERE

Litter Number _____

Farrowing interval or days to first farrowing _____
Date of last farrowing _____

171

Fig. 73. Pedigree record sheet

Meat and Livestock Commission—M.L.C.

The Meat and Livestock Commission was founded in 1968, and took over the Pig Industry Development Authority (known as P.I.D.A.) which was founded in 1958. During its ten-year life P.I.D.A. had a tremendous influence on the pig industry. It took over National Pig Recording and Progeny Testing, and established a National Feed Recording Scheme; sire performance testing, the Accredited Herd Scheme with 'élite' and accredited herds; pioneered artificial insemination in pigs, and maintained a research and advisory service for pig producers.

The M.L.C. is now continuing with the majority of the old P.I.D.A. schemes, but, quite naturally are making some changes in the way they serve the industry. For example, official sow and litter recording has ceased, but the food-recording scheme is still available to producers.

The Accreditation Scheme has been changed from 'élite' and accredited herds to nucleus and reserve nucleus herds and a third category known as multiplier herds.

The nucleus herds contain the very best pigs of high breeding potential. All the animals are rigorously tested and their performance must be of the highest possible standard. The reserve nucleus herd are also tested and may as a result of high performance be promoted into the Nucleus Herd Scheme.

Multiplier herds are those based on nucleus stock and who adopt 'on the farm recording', which means that the farmer tests his own stock rather than sending pigs to a National Performance Testing Centre. The M.L.C. assist the breeder by making available the use of ultrasonic testing equipment to assess the likely carcass quality of live animals and by advising the farmer on the compilation and interpretation of records of performance.

'On-the-farm testing' facilities are also available to commercial pig farmers wishing to improve breeding stock by the use of ultrasonics and keeping meticulous records as an aid to their normal selection programme.

The M.L.C. have in addition a number of valuable ancillary schemes which includes artificial insemination, premium boar scheme and feed recording.

The National Agriculture Centre—N.A.C.

The N.A.C. was founded by the Royal Agricultural Society of England in 1963 and formally opened in 1967. The Centre is situated in Warwickshire and provides agriculturalists with an enormous amount of information about present-day farming methods and likely future developments.

One of the main features at the Centre is a highly successful pig demonstration unit. Both pedigree and commercial cross-bred sows are kept under a variety of housing conditions to produce pork and bacon pigs. The visitor can study the various systems of management and is provided with comprehensive records of the pigs' performance. The housing accommodation includes sow stalls—sow yards—maternity farrowing—individual rearing and group suckling. A wide range of materials and equipment, with their costs, are on display.

A further aid to pig farmers is the Farm Buildings Centre which publishes information on the various types of pig housing and the materials used in their construction.

Twenty-two Physical and Financial Records

In order that the pig enterprise may be both profitable and efficient, it is essential that proper breeding records, and some form of cost accounts, are kept. This will enable the farmer to judge the performance of his pigs and provide valuable information when making future budgets.

Breeding records

The first essential is to be able to positively identify each sow and boar in the herd. With small herds the farmer will know each animal and probably give each sow a name, but in the larger herd a number should either be tattooed, notched, or tagged in the ear, or the body freeze branded.

A simple record card may then be used for each sow and such information as the date served, name of boar, date of farrowing, numbers born and reared, with weight of litter at three weeks and weaning, etc., recorded. This record of performance can then be used to compare with other sows, and obviously will be useful for culling those with poor performance, or when selecting future stock.

Alternatively, you can buy or build your own comprehensive wall chart, which if properly used and maintained will show at a glance the progress of every sow in the herd.

Livestock movement register

All livestock owners must, by law, keep a register of the movement of cattle, sheep, and pigs. This is necessary in order that the police and veterinary surgeons can trace livestock after they have left the farm should there be an outbreak of one of the

Sow Record Card

| Name or Number: | | Born: | | Sire: | | Dam: |

Litter	Name of Boar	Service Date	Date Farrowed	Number Born	Number Reared	3-week weight (lb)	Date weaned	Total weight (kg)	Average weight (kg)	REMARKS (e.g. Grading results)
1st										
2nd										
3rd										
4th										
5th										
6th										
7th										
8th										

Fig. 74. Sow record card

175

notifiable diseases, for example swine fever or foot-and-mouth disease.

The police have powers to call on a farmer and ask to see his movement book at any reasonable time.

Enterprise costs and budgeting

It is quite impossible to state in a book of this kind realistic figures for either costs or returns in pig production, because

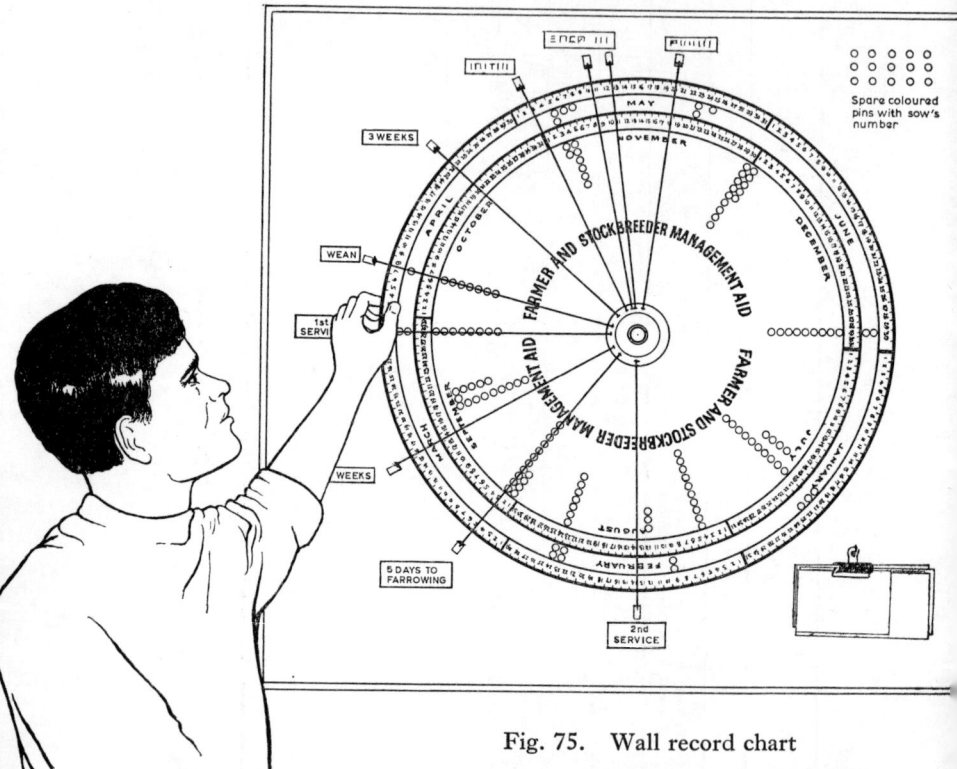

Fig. 75.   Wall record chart

they change rapidly, and what is true today may be quite different in a few months' time.

However, we shall discuss the various factors that contribute to financial success, but you must include your own figures, both financial and physical, for the conditions on your farm, at the time you prepare your budget.

176

Movement of Animals (Records) Order, 1960

Date of Movement	Particulars of each bovine animal moved to or moved from premises mentioned on front cover				Number of sheep, goats or pigs (specifying which)	Movements to premises mentioned on front cover	Movements from premises mentioned on front cover
	Breed	Age	Sex	Ear Mark or Ear Tag. No.		Movements from which moved (including Market, Saleyard or Fair) *and/or* Name and Address of person from whom delivery was taken	Premises to which moved (including Market, Saleyard or Fair) *and/or* Name and Address of person taking delivery.
						N.B.- Both of these particulars are to be entered if available	N.B. - Both of these particulars are to be entered if available

Fig. 76. Livestock movement register

177

Breeding pigs—factors affecting profitability

The number of weaners reared per sow per year
The weight of piglets at weaning
Annual cost of feeding the sow
Herd depreciation
Cost of housing
Labour and miscellaneous costs
Boar-service cost

The number of weaners reared per sow per year

The average number of pigs born per litter in the U.K. appears to be around 9–10, and mortality rate for suckling pigs is about 20%. Theoretically the sow should produce two or more litters per year, but in practice many herds average only 1·8. Thus the average number of pigs reared per sow per year is about 14–15. Every effort must be made to improve this number—the use of farrowing crates—warm, dry housing; early weaning at three weeks or six weeks are all ways in which production can be improved. But it is equally important that the 'extra' pigs reared are of good quality. It is no use rearing 'runt' pigs.

The weight of piglets at weaning

Pigs that have a good start in life and weigh 18–20 kg at eight weeks will usually 'grow on' with little difficulty and reach bacon weight in less time than pigs that have a poor start.

Annual cost of feeding the sow

The annual cost of feeding a sow may be easily checked, for example:

> Gestation period 114 days at 2·5–2·7 kg approx. 300 kg
> Suckling period, birth to weaning, say 56 days
>   at 5–6 kg = approx.   300 kg
> including creep feed
>
>   ───────
>
>   600 kg
>   ───────

With two litters a year this would mean an annual consumption of approximately 1·2 tons.

178

Herd depreciation

This is the difference between the purchase price and the sale price divided by the lifetime of the sow—for example:

> Purchase price for in-pig gilt say   £40–0
> Sale price after two years (four litters)   20–0

Depreciation of £20 over two years equals £10 per year or £5 per litter.

Obviously if you rear your own replacement stock you will only change your gilt at the cost of production, which in many cases may be equal to the sale price of the cull sow, and so you would have little or no depreciation.

Housing costs

This is a very difficult figure to arrive at—new farrowing houses are expensive to buy and to maintain and operate. If you pay £150–£200 per farrowing pen and write this off over ten years you will have an annual depreciation of £15–20. This must be divided by the number of farrowings that take place in the pen during the year—say five with no rest period. The cost per farrowing will be £3–£4 per litter or £6–£8 per sow per annum. In addition to this must be added dry sow accommodation.

On the other hand it may be that you have adapted old buildings into suitable sow housing for very little and your annual cost only amounts to a few pence per litter.

Labour costs

Where a full-time pigman is employed the labour cost is easily found by simply dividing the man's annual wages (including any bonus—national health, pension scheme, or tied cottage) by the number of sows in the herd, for example if one man looks after 100 sows and earns £2000 per annum, the cost per sow will be £20 per year.

Boar-service charge

This cost is calculated by adding together the annual food, labour, housing and depreciation cost, by the number of sows served by the boar, for example:

179

```
Cost of boar    £70
Sale price       10
                ───
                 60
                ───
```

```
                                          per annum
Average life of boar say 3 years      = £20
Food cost = approx. 1 tonne           =  40
Housing—vet.—miscellaneous, say =  10
Labour                                =  10
                                         ───
                                          80
                                         ───
```

If the boar serves forty sows per annum this amounts to £2 per sow or £1 per litter.

The total costs would, therefore, be approximately:

		A		B
Food—1–2 tonne	from	£50–0	to	£50–0
Depreciation	from	Nil	to	10–0
Labour	from	20–0	to	20–0
Housing	from	1–0	to	8–0
Boar service	from	2–0	to	3–0
Miscellaneous—light—water—telephone, etc.		1–0	to	2–0
		£74–0	to	£93–0

Our total costs must now be divided by the number of pigs reared per sow per annum:

Cost of weaner	A	B
14 pigs per year	£5·28	£6·64
16 pigs per year	4·62	5·81
18 pigs per year	4·11	5·16
20 pigs per year	3·70	4·65

We can now see the wide variation in the cost of producing weaner pigs ranging from as low as £3·70 up to £6·64 per pig.

Budgets for producing pork, bacon and heavy hogs are calculated in similar ways by estimating the cost of the weaner, food, labour, housing and miscellaneous, and deducting this from the estimated return per pig.

Twenty-three Health and Disease

'Health begins with Hygiene'

The pigman should be able to recognise quickly the ailing pig, and where necessary he must seek veterinary advice. In this last chapter we will look at the symptoms of some more common ailments and diseases.

Health has been discussed in several parts of this book, but to recapitulate: the healthy pig may be recognised by its alert appearance, eagerness to come forward to food, shiny coat, curly tail and weight for age. Its dung will be firm and slightly moist.

The sick pig is reluctant to eat, is usually lying down, and may be trembling or have skin discolouration. Sick pigs may become constipated or scour. Coughing, 'blowing', or a staggering gait may be observed. The pig's normal temperature is 38·9°C.

Notifiable diseases

Anthrax

All domestic animals, and man himself, are susceptible to this most deadly disease. Anthrax is caused by the microbe *bacillus anthracis*, which in its spore form is capable of living in the soil for many years.

In cattle and sheep the disease is extremely acute. The animal may be inspected and appear perfectly healthy, only to be found a few hours later dead. Therefore, any sudden deaths should be considered as possible anthrax, and the police or your veterinary surgeon informed.

With pigs, the disease is usually less acute. The germ attacks the throat region, causing severe inflammation, and choking, and the disease may run from 8–16 hours before death occurs.

It is most important that the disease is properly diagnosed by a veterinary surgeon, who will take a blood sample from the corpse. On no account should you try to do a post mortem, since infection is extremely dangerous and easily picked up.

Swine fever

Swine fever is an infectious, notifiable disease which only affects pigs, usually those kept indoors. It is caused by a virus which is transmitted from infected to healthy pigs in the dung and urine.

Symptoms are extremely varied. The pigs may go off their food and show a high temperature; later they will lie down in a dark corner of the sty; if moved, they will squeal and may show weakness in their hind legs; some may cough, and there may be scouring.

If these symptoms occur, the stockman should lose no time in telephoning his veterinary surgeon and discussing the symptoms with him or notify the police. It is possible to vaccinate pigs against swine fever. Consult your veterinary surgeon.

Foot-and-mouth disease

Foot-and-mouth disease is perhaps the most well-known disease that affects stock, following the tragic outbreaks of 1967–8 when the entire British Isles came under supervision.

The disease is caused by a virus which may be transmitted by direct contact or through uncooked meat, hides, bones, feedingstuffs, vehicles, personal clothing and vermin. Migratory birds, on their flights from country to country, may also be carriers.

Symptoms

The disease can affect all cloven-hooved livestock. At first, the animals lose their appetite, become dull and sluggish. Lactating animals will suddenly lose their milk yield and there may be lesions around the udder.

With cattle, affected stock will smack their lips, and long strings of saliva will hang from the mouth.

182

On examination, blisters will be found around the dental pad. Later, the blister will burst, thus liberating the infective virus. Lameness will also occur, and blisters will be found in the feet.

With pigs, the mouth lesions are not as common as in cattle, but foot lesions are more common and will cause the pig considerable pain, making it difficult to stand. Pigs will be most distressed and found squealing.

Needless to say, the earlier the Ministry of Agriculture veterinary officer is called in, the better. If the disease is confirmed, all the cattle and sheep and pigs on the farm will be slaughtered and compensation paid to the owner.

Parasites affecting pigs

Roundworms affect pigs of all ages, but are most troublesome in young pigs, especially at weaning time. The most common worm is *Ascaris lumbricoides*, the large roundworm, but recent evidence suggests that other worms, particularly the nodular worms (*Oesophagostomum spp*), the stomach worm (*Hyostrongylus rubidus*) and the whipworms (*Trichuris suis*) are causing trouble, especially in breeding sows.

Life-cycle (ascaris)

The adult worms live in the intestines, where the female worms lay their eggs. These pass on to the ground in the droppings. On the ground, the eggs become *infective* after a few weeks. The eggs have a remarkable vitality and may keep alive for up to five years.

If picked up, the eggs hatch into larvae, which bore through the stomach wall and enter the bloodstream. They are then carried in the blood to the lungs; here the larvae do considerable damage to lung tissue and cause the pig to cough.

The larvae then crawl up the trachea, or may be coughed up into the mouth. They then pass down the oesophagus (gullet), enter the intestine, and develop into mature worms.

The adult worm lives on the food in the intestines, and also sucks blood out of the intestinal wall, thus robbing the pig of nourishment, and this leads to loss of body condition and to anaemia.

Symptoms are general unthriftiness, a staring coat, straight tails, stunted growth and frequent coughing.

183

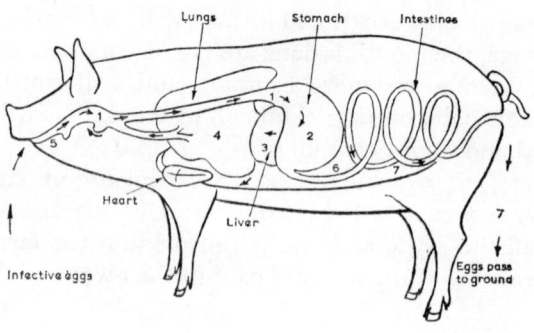

Fig. 77.   Life-cycle of the roundworm

Control

Do seall weaners when eight to ten weeks old with a safe anthelmintic, such as Piperazine, to control ascaris worms.

Sows should be dosed about two weeks before farrowing, with a wide anthelmintic to control roundworms, nodular worms and whipworms.

Prevention

Wash the sow's udder with warm water and a mild antiseptic just before farrowing to kill any worm eggs adhering to her sides.

Disinfect dunging passage in bacon pens weekly to kill worm eggs before they become infective.

Lice

Lice are a common problem with pigs of all ages. They crawl over the skin and suck blood, which, of course, causes severe irritation. In extreme cases young pigs have been known to die from loss of blood. The lice are usually found near the ears, inside the elbows and on the belly.

184

Treatment

Fortunately, lice are very easy to kill. Gammexane, derris, or D.D.T. will give complete control.

You may also use waste oil, which, if spread along the back, will kill the lice, but unfortunately will not control any unhatched eggs. It is necessary, therefore, to repeat the treatment in 7–10 days.

Mange

Mange in pigs is caused by a blood-sucking parasite that burrows under the skin. It sets up intense irritation and causes scabs to break out. Affected pigs rapidly lose condition if left untreated.

Treatment

Bathe the infected areas with warm water, and dress with a gammexane dressing, available from your veterinary chemist.

Bacterial diseases

Agalactia

This means the absence of milk in the udder after parturition. In sows, this can take many forms. The udder may become extremely hot and painful or very cold and hard. It may become swollen, so that the quarters are irregular in size. There may be a complete absence of milk or an abundance of bloodstained or otherwise discoloured milk.

The piglets must be given supplementary food, either proprietary sow milk replacement or cow's milk.

Make the sow as comfortable as possible, and summon your veterinary surgeon immediately.

Scouring in young pigs—E. Coli

Profuse diarrhoea or scours may occur in young pigs from about 3 days up to 10–12 weeks. The infection is known to be caused by a bacterium called *Escherichia Coli* or *E. Coli*.

There are at present over 120 strains of *E. Coli* known to

scientists. Some strains are controlled by antibiotics, whilst others appear to be unaffected, and the scouring will continue despite treatment.

Prevention

Strict attention to hygiene is the best way to control this disease. Scrub each farrowing pen with hot water and washing soda after each sow and litter is removed. Try to rest the pen for eight weeks between farrowing.

Necrotic enteritis

This can affect growing pigs from around weaning time until they reach sixteen weeks. The disease is recognised by a progressive unthriftiness, loss of condition and the appearance of a dirty, greasy, brown skin. The pigs stand with an arched back and scour profusely.

The disease is caused by a bacterium called *Salmonella suipester*, which lives in the pig's bowel.

Prevention

The best is good housing, providing warm, draught-free conditions, kept scrupulously clean. Feed the pigs balanced rations containing adequate vitamins.

Treatment

Feeding high-level antibiotics has given reasonable results but is not always a complete cure.

Swine erysipelas—'purples' or 'diamonds'

This is caused by the bacterium *Erysipelothrix rhusiopathiae*.

Symptoms

It is characterised by reddish or purple square patches on the skin: hence the name 'purples' or 'diamonds'. The disease may affect pigs in three ways: acute, sub-acute and chronic.

186

Acute form is most common in very hot weather. The pig goes off its food and runs a very high temperature of around 41°C, the purple patches soon occurring. Prompt treatment by a veterinary surgeon is necessary or death will occur. Treatment is usually injection of antibiotics.

Sub-acute is the commonest form of erysipelas. The pigs go lame and there may be swelling of the knees and hocks. Purple patches occur about the second or third day of the attack, and then disappear. If treated in good time the pigs will make quite a spectacular recovery.

Chronic often occurs in pigs that have recovered from the acute or sub-acute attack. The pigs will eat their food but fail to thrive satisfactorily. Breathing may become difficult and coughing develop. This is because the erysipelas germs have affected the heart valve during the acute attack. Unfortunately, there is no cure, and the pig will eventually die.

Prevention

All breeding stock should be vaccinated against the disease.

Baby pig disease

This may develop in pigs when 2–3 days old. It is caused by a lack of blood sugar. The baby piglets refuse to suckle, they become very weak, and will die if not treated quickly.

Treatment

Remove the piglets from the sow, and bottle feed with a proprietary sow-milk substitute. Keep the piglets warm by placing them under an infra-ray lamp.

Your veterinary surgeon should treat the sow.

Enzootic pneumonia (formerly known as virus pneumonia or V.P.P.)

According to recent reports, it is estimated that between 40–50% of the total pig population is affected with enzootic pneumonia. The disease causes a mild coughing (rather similar to mild bronchitis in humans) which slows down the animals' progress. Food conversion will widen, and the pigs may take a little longer to reach slaughter weight.

Most farmers look upon the disease as something to live with, and claim that with reasonable management they can get by. However, there is an increasing number of people who are keeping enzootic pneumonia-free herds, and they claim that the improved efficiency of these pigs more than justifies the cost of establishing such a herd.

Symptoms

In suckling pigs the symptoms may be very mild indeed, but after weaning some coughing will be noticed, with perhaps sneezing and occasionally *lacrymatia* (running eyes).

It is possible for older pigs to have acute form, with the odd death, but this is very rare.

The chronic condition will continue until the pig is slaughtered (at 6–7 months), and examination at the slaughterhouse will generally show solidification in the front lobes of the lungs.

The carcass is not affected and is quite suitable for human consumption.

Control

It may be possible to select pigs in your herd that are free from the infection and to breed replacement stock from them. This is done by asking the abattoir personnel to examine the lungs of pigs slaughtered out of sows that are believed to be clean. If the lungs of the hogs are clear, then litter sister gilts may be bred from. It is most important that clean pigs are kept well away from the infected ones until you build up sufficient numbers to be entirely free.

It is probably better (but it takes a great deal of courage) to sell out all the pigs, thoroughly clean the housing, and buy in fresh stock from an enzootic-free herd.

Bowel oedema

This is usually associated with weaner pigs that have been subjected to a sudden change of environment or change of diet.

It may, however, affect feeding pigs up to four or five months old.

Pigs suffering with oedema may throw a fit or be seen staggering around the pen with little or no control over their movements. The eyelids may be swollen and the face is puffy. Some of the pigs may die, whilst the others may make a quick recovery.

Prevention

Always make changes in the pigs' rations gradually and if you buy in-store pigs from market feed them lightly for a day or so until they settle in. A dose of castor oil or Epsom salts will be beneficial.

Treatment

Call in your veterinary surgeon immediately. He will probably inject all the surviving pigs with an antibiotic and prescribe a suitable purgative—50 g Epsom salts per weaner is often used.

T.G.E.—transmissible gastro-enteritis

The disease known as T.G.E. has caused considerable losses in breeding herds in recent years. It is a highly infectious virus disease and appears to attack in yearly cycles. In 1969–70 there were nearly 600 cases reported in the U.K.

Symptoms

The disease attacks suckling pigs, usually between one and three weeks old. The disease takes the form of acute gastro-enteritis; there is considerable scouring, which is soon followed by death. Being highly infectious, the disease can wipe out several litters if the pigs are housed in communal quarters.

Treatment

At the time of writing there is no satisfactory treatment, although there are several vaccines on trial. The owner should lose no time in calling in his veterinary surgeon in order to help prevent the disease from spreading.

Prevention

It would appear that the disease is carried from one farm to another by 'man and machine', rather than being airborne. This means that hygienic conditions must be maintained at all times. When visiting other pig farms it is much better not to take your working clothes or gumboots—as this can be a means of carrying T.G.E. and other sources of infection.

An advantage is to house the sows separately. If you do batch farrow in a communal house, then make some provision for isolating a sow and litter should you have a suspected case.

Disease control

1 Install a footbath containing a suitable disinfectant for all visitors to dip their footwear. Ask visitors to leave their gumboots at home!

2 Isolate all bought-in pigs for at least one month before allowing them to come in contact with the main herd.

3 Avoid feedingstuffs and pig-collection lorries from coming close to the pig buildings. If possible build a pig-loading bay some distance from the pens and walk the pigs to the bay through a 'race'.

4 Thoroughly clean and disinfect all buildings immediately after use. Scrape dung off the walls—then soak and scrub with hot water and either a detergent or washing soda.

5 Try to rest pens for two or more weeks between batches of pigs.

Fig. 78. Dipping boots

190

Conclusion

Healthy pigs are undoubtedly profitable pigs. This chapter is not a veterinary manual, but simply tries to help the pigman to be aware of the problems associated with keeping pigs. If it enables you to recognise the early signs of ill-health in your pigs, then it has achieved its objective. Always remember that there is no substitute for good stockmanship and attention to detail is always of vital importance.

APPROXIMATE
CONVERSION TABLES—READY RECKONER

Weight

| kilogram to lb | | lb to kilogram | | kilogram to cwt | | cwt to kilogram | |
kg	lb	lb	kg	kg	cwt	cwt	kg
0·5 =	1·10	0·5 =	0·23	25 =	0·49	½ =	25·4
1 =	2·20	1 =	0·45	50 =	0·98	1 =	51
2 =	4·41	2 =	0·9	100 =	1·97	2 =	101
3 =	6·61	3 =	1·36	150 =	2·95	3 =	152
4 =	8·82	4 =	1·8	200 =	3·94	4 =	202
5 =	11·00	5 =	2·26	250 =	4·92	5 =	254
6 =	13·20	6 =	2·7	300 =	5·90	6 =	305
7 =	15·40	7 =	3·1	350 =	6·89	7 =	355
8 =	17·60	8 =	3·6	400 =	7·87	8 =	404
9 =	19·80	9 =	4·0	450 =	8·86	9 =	456
10 =	22·0	10 =	4·5	500 =	9·84	10 =	508

Area

| Hectares to acres | | | Acres to hectares | | |
ha		acres	acres		ha
0·5	=	1·24	0·5	=	0·2
1	=	2·47	1	=	0·4
2	=	4·94	2	=	0·8
3	=	7·41	3	=	1·2
4	=	9·88	4	=	1·6
5	=	12·36	5	=	2·0
10	=	24·71	10	=	4·0

| Square metres to square feet | | | Square feet to square metres | | |
m^2		ft^2	ft^2		m^2
1	=	10·76	1	=	0·09
2	=	21·53	2	=	0·18
3	=	32·29	3	=	0·27
4	=	43·06	4	=	0·36
5	=	53·82	5	=	0·46
10	=	107·62	10	=	0·92

Length

Millimeters to inches			Inches to millimetres		
mm		inch	inch		mm
25	=	0·99	1	=	25·4
50	=	1·97	2	=	50·8
100	=	3·94	3	=	76·2
200	=	7·87	4	=	102·0
300	=	11·80	5	=	127·0
400	=	15·70	6	=	152·0
500	=	19·70	12	=	305·0

Metres to yards			Yards to metres		
m		yard	yard		m
0·5	=	0·55	0·5	=	0·46
1	=	1·09	1	=	0·91
2	=	2·19	2	=	1·8
3	=	3·28	3	=	2·7
4	=	4·37	4	=	3·6
5	=	5·47	5	=	4·5
10	=	10·90	10	=	9·1

Capacity

Litres to gallons			Gallons to litres		
litre		gallon	gallon		litre
0·5	=	0·11	0·5	=	2·27
1	=	0·22	1	=	4·55
2	=	0·44	2	=	9·09
3	=	0·66	3	=	13·64
4	=	0·88	4	=	18·18
5	=	1·10	5	=	22·73
10	=	2·20	10	=	45·46

Glossary of Terms

Pigs

Breeding stock

Boar	A mature entire male kept for breeding purposes
Sow	A mature female pig kept for breeding and suckling pigs
In-pig sow	Sows that are pregnant—usually called 'in pig' six weeks after service
Barren sow or 'empty sow'	Mature female pig that is not pregnant or suckling her litter. Since a barren sow is unprofitable, she should be fattened up and slaughtered
Suckling sow	Sow in lactation. Usually sows milk for six to eight weeks before the pigs are weaned
Gilt or hilt	A female pig
Breeding gilt	A young female pig that is intended for breeding
Maiden gilt	Female pig 6–8 months old
Served gilt	Female pig, approximately eight months old that has been mated to a boar
In-pig gilt	Provided the gilt does not come 'on heat' for six weeks after service she is considered to be pregnant and called in pig
Pure bred	Mother and father are of the same breed

Hybrid pigs	Hybrids are produced by selective crossing of certain strains, often using three or more breeds. The hybrid is claimed to have hybrid vigour

Fattening stock

Hog	Castrated male pig intended for slaughter as either pork, cutter, bacon or heavy hog

	Approx. age at slaughter	Liveweight	Dressing %
Pork	14–16 weeks	45– 55 kg	70–73 %
Cutter	16–20 weeks	70– 80 kg	72–74 %
Bacon	22–26 weeks	85– 95 kg	73–75 %
Heavy hog	28–32 weeks	115–130 kg	75–80 %

Clean pigs	Castrated males and unserved gilts which are healthy and intended for slaughter
Fat sow	Female pig that has bred pigs, but is now barren and in a suitable condition for slaughter

Terms used in pig production

Accredited	Breeding stock of superior quality —supported with records of performance
Amino-acids	Chemical substances found in protein foods, e.g. Lysine
Anaesthetic	A chemical used by veterinary surgeons to produce insensibility in stock when performing surgical operations
Anaemia	Illness caused by the lack of haemoglobin in the blood. Common in baby pigs
Anaerobic	Usually refers to bacteria that live in the soil without the use of oxygen
Artery	A blood vessel that carries blood away from the heart to all parts of the body

Artificial insemination, A.I.	A method of breeding widely used with pigs. Semen is collected from the male and inseminated into the female uterus with a rubber catheter
Ash	The mineral matter or 'minerals' found in feeding stuffs
Bacon	The cured meat from a pig carcass between 61 and 77 kg
Backfat	The layer of fat covering the muscle over the back and loin
Bloom	Sleek or shiny coat—indicates a healthy pig
'Blue' pig	Result of crossing a white breed with a black breed
Brawn	Manufactured by boiling the meat off the pig's head, trotters and tail.
Carcass or carcase	The dressed body after removal of hair and internal organs
C and K measurements	A measurement of backfat taken in millimetres with an optical probe (intrascope) at two fixed points. May also be assessed on live animals by the use of ultrasonic tests
Calorie	The amount of heat required to raise the temperature of 1 g of water from 14·5 to 15·5°C
Catheter	A long rubber tube simulating the boar's penis—used in artificial insemination
Carbohydrate	The elements, carbon, hydrogen and oxygen which combine together to form sugars, starch and fibre—carbohydrates are used to produce energy
Castration	Surgical removal of testicles
Condition	Indicates the degree of finish or fatness
Creep	A small soundly constructed pen where baby piglets can seek warmth and food and protection from overlaying by a clumsy sow

Cross-bred	The mating of two different breeds
Dam	The maternal parent
Dressed weight dead weight d.w.	The weight of carcass usually compared with liveweight to give dressing percentage
Dressing percentage k.o.% carcass yield	$\dfrac{\text{Carcass weight} \times 100}{\text{Liveweight}} = \dfrac{\text{dressing}}{\text{percentage}}$
d.l.w.g.	Daily liveweight gain
Digestible energy	The gross energy (or heat of combustion) of a food minus the gross energy of the corresponding faeces
d.w.	dead weight
Embryo	Developing pig in uterus
Farrowing or Parturition	Birth of a litter
Fecundity	A sow that breeds and rears better than average size litters is said to have good fecundity
Food conversion ratio	The number of kilograms of food required to increase the liveweight by one kilogram
Gammon or ham	The hind leg—the most valuable part of the carcass
Gestation	The time from conception to birth
Gestation period	114 days or 3 months 3 weeks 3 days
Heat period or 'on heat'	The period of oestrum when a sow or gilt will mate with the boar
Hybrid vigour	The superior performance of the offspring over the average performance of both parents, e.g. increase in litter size when two different breeds are mated together
Intrascope or optical probe	Instrument used for measuring the amount of backfat in the carcass
Kibble	Coarsely ground or broken grain
Litter	The pigs produced as the result of one complete farrowing—litter of pigs

Litter	Short straw used for bedding
Lot or pen of pigs	A single or group of pigs offered for sale by auction
l.w.	liveweight
Make-up	Deficiency payment for clean pigs eligible for the Government guaranteed price fatstock scheme
Maturity	Fully grown pig, i.e. complete development of bone, muscle and fat
Management	The skill and expertise of the farmer and pigman—includes the feeding—breeding—housing and day-to-day work of running a pig unit
Meal	Dry food, for example cereals, ground finely before feeding to the pigs
Meat	The flesh or muscle and fat of animals slaughtered
M.L.C.	Abbreviation for the Meat and Livestock Commission
Muscle	The red meat or lean meat cut from a carcass
Multiplier breeder	Farmer who increases or multiplies hybrid or pure-bred breeding stock for sale to commercial farmers
Nucleus breeder	Small number of breeders who keep very high-quality pedigree and recorded stock for sale to multiplier breeders
Performance testing	Measuring the performance of individual animals and comparing this with the performance of their contemporaries
Pedigree	Record of ancestry—usually shown as: Name and no. { Sire—name and no. Dam—name and no.
Progeny testing	Recording the performance of a boar or sows progeny

Pork	Fresh pigmeat (may also be frozen)
Puberty	The age when a boar or gilt becomes sexually mature i.e. capable of breeding
Placenta	The afterbirth or cleansing—membranes that surround the pigs when in uterus—removed after farrowing
Rectum	The final part of the bowel
Runt	Small, poorly developed pig—'runt' of the litter
Rig	Male with one undescended testicle
Rupture or scrotal hernia	Part of the intestine forced into the scrotum
Scrotum	Purse or bag containing the testicles of male animal
Semen	A fluid produced by the male, containing sperms
Service	The act of mating or copulation
Sow stall	Small pen in which pregnant sows are kept—sow allowed to stand up or lie down—cannot turn round
Seedy cut	Discolouration of the meat in bacon carcass—due to pigmentation of the mammary gland; this is only likely to occur with coloured breeds
Sound soundness	Healthy animals suitable for breeding or meat production—free from serious fault
Sire	Male—paternal father
Spay	The removal of the ovaries from gilts to prevent breeding (rarely done today)
Testicles	Male reproductive organs which produce sperms and testosterone
Testosterone	Male hormone responsible for secondary male characteristics—masculine appearance, desire to mate, etc.

Tethered sows	Similar to sow stalls. Sows tethered with chain or strap around neck and confined in a small space during their pregnancy. Also, sows are sometimes tethered with a harness and long chain when kept outdoors
Weaner pool	A fairly large strawed yard with a sleeping kennel area where around fifty or so weaners are kept from 18–20 kg liveweight until they reach about 50 kg

Selected Bibliography

MINISTRY OF AGRICULTURE, FISHERIES AND FOOD, *Fatstock Guarantee Scheme 1971–72,* London, H.M.S.O. (1971).

BUCKETT, M., *Introduction to Livestock Husbandry,* Oxford, Pergamon Press (1965).

BRITISH VETERINARY ASSOCIATION, *Handbook on Meat Inspection,* London, British Veterinary Association (1965).

HAMMOND, J., *Progress in the Physiology of Farm Animals,* London, Butterworths, **1** (1954), **2** (1955), **3** (1959).

JENNINGS, J., *Feeding, Digestion and Assimilation in Farm Animals,* Oxford, Pergamon Press (1965).

JONES, E., *'Just Your Meat' or the Judging of Meat Animals* (2nd ed.), London, Headley (1955).

LAWRIE, R., *Meat Science,* Oxford, Pergamon Press (1966).

LINE, E., *The Science of Meat and Biology of Food Animals,* London, Meat Trades Journal, 2 volumes (1932).

McMEEKAN, C. P., *The Principles of animal production* (2nd ed.), London, Whitcombe and Tombs (1934).

MARSHALL, F. H. and HALMAN, E. T., *The Physiology of Farm Animals* (4th ed.), Cambridge University Press (1948).

PARK, R. D., *Animal Husbandry,* London, Oxford University Press (1961).

STUBBS, D. R. and CATO, C. A., *Know Your Farm Stock,* Aberdeen, Scottish Association Y.F.C.

THOMAS, D. G. M. and DAVIES, W. I. J., *Animal Husbandry,* London, Cassell (1963).

WILLIAMSON, G. and PAYNE, W. J. A., *Animal Husbandry* (2nd ed.), London, Longmans (1965).

AGRICULTURAL RESEARCH COUNCIL, *The Nutrient Requirements of Farm Livestock,* London, Agricultural Research Council 3, Pigs, technical reviews and summaries (1967).

AGRICULTURAL RESEARCH COUNCIL, *The Nutrient Requirements of Farm Livestock,* London, H.M.S.O. 3, Pigs, summaries of estimated requirements (1966).

HAMMOND, J., *Farm Animals; their Breeding, Growth and Inheritance,* (3rd ed.), London, Edward Arnold (1960).

201

PAWSON, H. C., *Robert Bakewell; Pioneer Livestock Breeder,* London, Crosby Lockwood (1957).

HAFEZ, E., S., E., *Reproduction in Farm Animals* (2nd ed.), Philadelphia, Lea and Febiger (1968).

HAMMOND, Sir John, *Animal breeding,* London, Edward Arnold (1963).

MINISTRY OF AGRICULTURE, *Rations for Livestock,* R. E. Evans (15th ed.), London, H.M.S.O. (1960).

DEPARTMENT OF AGRICULTURE AND FISHERIES OF SCOTLAND, *The Feeding of Farm Animals,* Edinburgh, H.M.S.O. (1966).

NELSON, R. H., *An Introduction to Feeding Farm Livestock,* Oxford, Pergamon Press (1964).

TYLER, C., *Animal Nutrition* (2nd ed.), London, Chapman and Hall (1964).

HALNAN, E. T., GARNER, F. H., EDEN, A. *The Principles and Practice of Feeding Farm Animals* (5th ed.), London Estates Gazette.

SHEEHY, E. J., *Animal Nutrition,* London, MacMillan (1955).

HAGEDOORN, A. L., *Animal Breeding* (6th ed.), London, Crosby Lockwood (1962).

BARRON, N., *The Pig Farmer's Veterinary Book,* Ipswich, Dairy Farmer Books (1957).

MILLER, W. C., *Veterinary Dictionary* (8th ed.), London, Black and Black (1967).

MERICK VETERINARY MANUAL, *A Handbook of Diagnosis and Therapy for the Veterinarian* (2nd ed.), Rachway, N.J., U.S.A., Merick (1961).

PATERSON, J. D., *Good and Healthy Animals,* London, Hodder and Stoughton (1947).

ANTHONY, D. J., *Diseases of the Pig and its Husbandry,* London, Baillière, Tindall and Cox (1940).

UNIVERSITY OF BRISTOL, University Department of Economics (Agricultural Economics), *An economic study of pig production in South West England,* Estelle Burnside and R. C. Rickard (1961).

DAVIDSON, H. R., *The Production and Marketing of Pigs* (3rd ed.), London, Longmans (1966).

FISHWICK, V. C., *Pigs, their Breeding and Management* (7th ed.), London, Crosby Lockwood (1959).

BARRON, N., *The Pig Farmer's Veterinary Book* (4th rev. ed.), Ipswich, Farming Press (Books) (1964).

BARRON, N., *The Pig Farmer's Veterinary Book* (2nd rev. ed.), Ipswich, Dairy Farmer (Books) (1957).

CAMBRIDGE, University School of Agriculture Farm Economics Branch, *A Comparison of Pig Production in England, Denmark and Holland,* by F. G. Sturrock and R. F. Ridgeon, Cambridge (1966).

CHAPPELL, C., *Keeping Pigs,* Hart-Davis (1953).

COEY, W. E., *A Modern Guide to Pig Husbandry,* London, Vinton (1958).

FITZHUGH, J., *Pig Breeding,* London, Land Books (1961).

202

MINISTRY OF AGRICULTURE, FISHERIES AND FOOD, *Diseases of Pigs,* H. I. Field, (2nd ed.), London, H.M.S.O. (1964).

MINISTRY OF AGRICULTURE, FISHERIES AND FOOD, *The Farm as a Business,* London, H.M.S.O. Part 4: Aids to management, pigs (1963).

MINISTRY OF AGRICULTURE, FISHERIES AND FOOD, *Housing the Pig,* Bulletin 160 (2nd ed.), London, H.M.S.O. (1962).

MINISTRY OF AGRICULTURE, FISHERIES AND FOOD, *Pig Feeding and Management,* London, H.M.S.O. (1963).

LUSCOMBE, J., *Pig Husbandry* (2nd rev. ed.), Ipswich, Farming Press (1970).

LUSCOMBE, J., *Making Money from Pigs,* Ipswich, Dairy Farmer.

PRICE, W. T., *The Pig: Modern Husbandry and Marketing,* London, Geoffrey Bles (1962).

SAINBURY, D., *Pig Housing* (2nd ed.), Ipswich, Farming Press, (1970).

BARRON, N., *The Pig Farmer's Veterinary Book: a complete guide to the farm treatment of pig diseases,* Farming Press, Ipswich.

Monthly

Pig Farming, Farming Press, Lloyds Chambers, Ipswich.